CBEST Math in 10 Days!

The Most Effective CBEST Math Crash Course

By

Reza Nazari

Copyright © 2023
Effortless Math Education Inc.

All rights reserved. No part of this publication may be reproduced, stored in a retrieval system, or transmitted in any form or by any means, electronic, mechanical, photocopying, recording, scanning, or otherwise, except as permitted under Section 107 or 108 of the 1976 United States Copyright Ac, without permission of the author.

Effortless Math provides unofficial test prep products for a variety of tests and exams. It is not affiliated with or endorsed by any official organizations.

All inquiries should be addressed to:
info@effortlessMath.com
www.EffortlessMath.com

ISBN: 978-1-63719-602-1

Published by: **Effortless Math Education Inc.**

For Online Math Practice Visit www.EffortlessMath.com

Welcome to
CBEST Math Prep 2024

Thank you for choosing Effortless Math for your CBEST Math test preparation and congratulations on making the decision to take the CBEST test! It's a remarkable move you are taking, one that shouldn't be diminished in any capacity. That's why you need to use every tool possible to ensure you succeed on the test with the highest possible score, and this extensive study guide is one such tool.

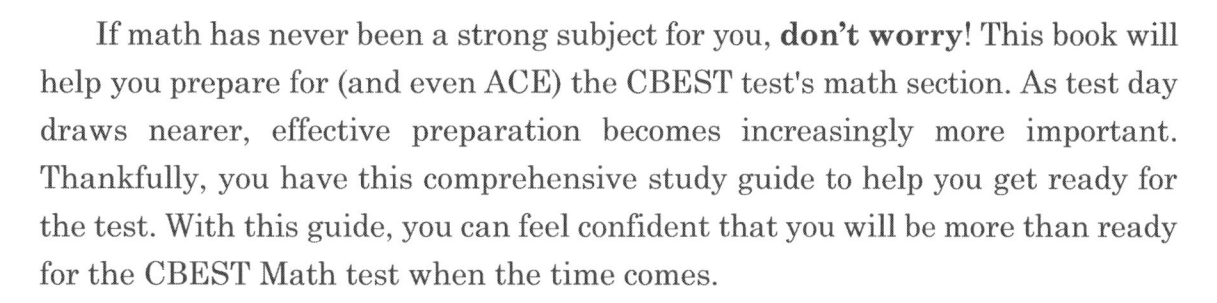

If math has never been a strong subject for you, **don't worry**! This book will help you prepare for (and even ACE) the CBEST test's math section. As test day draws nearer, effective preparation becomes increasingly more important. Thankfully, you have this comprehensive study guide to help you get ready for the test. With this guide, you can feel confident that you will be more than ready for the CBEST Math test when the time comes.

First and foremost, it is important to note that this book is a study guide and not a textbook. It is best read from cover to cover. Every lesson of this "self-guided math book" was carefully developed to ensure that you are making the most effective use of your time while preparing for the test. This up-to-date guide reflects the 2024 test guidelines and will put you on the right track to hone your math skills, overcome exam anxiety, and boost your confidence, so that you can have your best to succeed on the CBEST Math test.

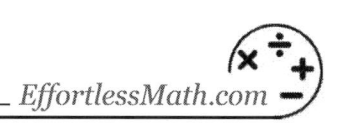

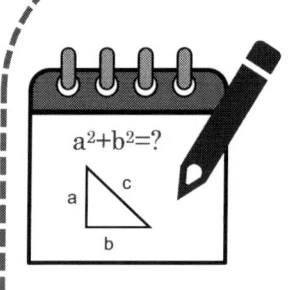

This study guide will:

☑ Explain the format of the CBEST Math test.

☑ Describe specific test-taking strategies that you can use on the test.

☑ Provide CBEST Math test-taking tips.

☑ Review all CBEST Math concepts and topics you will be tested on.

☑ Help you identify the areas in which you need to concentrate your study time.

☑ Offer exercises that help you develop the basic math skills you will learn in each section.

☑ Give **2 realistic and full-length practice tests** (featuring new question types) with detailed answers to help you measure your exam readiness and build confidence.

This resource contains everything you will ever need to succeed on the CBEST Math test. You'll get in-depth instructions on every math topic as well as tips and techniques on how to answer each question type. You'll also get plenty of practice questions to boost your test-taking confidence.

In addition, in the following pages you'll find:

➤ **How to Use This Book Effectively** – This section provides you with step-by-step instructions on how to get the most out of this comprehensive study guide.

➤ **How to study for the CBEST Math Test** – A six-step study program has been developed to help you make the best use of this book and prepare for your CBEST Math test. Here you'll find tips and strategies to guide your study program and help you understand CBEST Math and how to ace the test.

- ➢ **CBEST Math Review** – Learn everything you need to know about the CBEST Math test.

- ➢ **CBEST Math Test-Taking Strategies** – Learn how to effectively put these recommended test-taking techniques into use for improving your CBEST Math score.

- ➢ **Test Day Tips** – Review these tips to make sure you will do your best when the big day comes.

Effortless Math's CBEST Online Center

Effortless Math Online CBEST Center offers a complete study program, including the following:

- ✓ Step-by-step instructions on how to prepare for the CBEST Math test
- ✓ Numerous CBEST Math worksheets to help you measure your math skills
- ✓ Complete list of CBEST Math formulas
- ✓ Video lessons for all CBEST Core Math topics
- ✓ Full-length CBEST Math practice tests
- ✓ And much more…

No Registration Required.

Visit **EffortlessMath.com/CBEST** to find your online CBEST Math resources.

How to Use This Book Effectively

Look no further when you need a study guide to improve your math skills to succeed on the math portion of the CBEST test. Each chapter of this comprehensive guide to the CBEST Math will provide you with the knowledge, tools, and understanding needed for every topic covered on the test.

It's imperative that you understand each topic before moving onto another one, as that's the way to guarantee your success. Each chapter provides you with examples and a step-by-step guide of every concept to better understand the content that will be on the test. To get the best possible results from this book:

- **Begin studying long before your test date.** This provides you ample time to learn the different math concepts. The earlier you begin studying for the test, the sharper your skills will be. Do not procrastinate! Provide yourself with plenty of time to learn the concepts and feel comfortable that you understand them when your test date arrives.
- **Practice consistently.** Study CBEST Math concepts at least 20 to 30 minutes a day. Remember, slow and steady wins the race, which can be applied to preparing for the CBEST Math test. Instead of cramming to tackle everything at once, be patient and learn the math topics in short bursts.
- Whenever you get a math problem wrong, **mark it off, and review it later** to make sure you understand the concept.
- Start each session by **looking over the previous material.**
- Once you've reviewed the book's lessons, **take a practice test at the back of the book** to gauge your level of readiness. Then, review your results. Read detailed answers and solutions for each question you missed.
- **Take another practice test** to get an idea of how ready you are to take the actual exam. Taking the practice tests will give you the confidence you need on test day. Simulate the CBEST testing environment by sitting in a quiet room free from distraction. Make sure to clock yourself with a timer.

EffortlessMath.com

How to Study for the CBEST Math Test

Studying for the CBEST Math test can be a really daunting and boring task. What's the best way to go about it? Is there a certain study method that works better than others? Well, studying for the CBEST Math can be done effectively. The following six-step program has been designed to make preparing for the CBEST Math test more efficient and less overwhelming.

Step 1 - Create a study plan
Step 2 - Choose your study resources
Step 3 - Review, Learn, Practice
Step 4 - Learn and practice test-taking strategies
Step 5 - Learn the CBEST Test format and take practice tests
Step 6 - Analyze your performance

STEP 1: Create a Study Plan

It's always easier to get things done when you have a plan. Creating a study plan for the CBEST Math test can help you to stay on track with your studies. It's important to sit down and prepare a study plan with what works with your life, work, and any other obligations you may have. Devote enough time each day to studying. It's also a great idea to break down each section of the exam into blocks and study one concept at a time.

It's important to understand that there is no "right" way to create a study plan. Your study plan will be personalized based on your specific needs and learning style.

Follow these guidelines to create an effective study plan for your CBEST Math test:

★ **Analyze your learning style and study habits** – Everyone has a different learning style. It is essential to embrace your individuality and the unique way you learn. Think about what works and what doesn't work for you. Do you prefer CBEST Math prep books or a combination of textbooks and video lessons? Does it work better for you if you study every night for thirty minutes or is it more effective to study in the morning before going to work?

★ **Evaluate your schedule** – Review your current schedule and find out how much time you can consistently devote to CBEST Math study.

★ **Develop a schedule** – Now it's time to add your study schedule to your calendar like any other obligation. Schedule time for study, practice, and review. Plan out which topic you will study on which day to ensure that you're devoting enough time to each concept. Develop a study plan that is mindful, realistic, and flexible.

★ **Stick to your schedule** – A study plan is only effective when it is followed consistently. You should try to develop a study plan that you can follow for the length of your study program.

★ **Evaluate your study plan and adjust as needed** – Sometimes you need to adjust your plan when you have new commitments. Check in with yourself regularly to make sure that you're not falling behind in your study plan. Remember, the most important thing is sticking to your plan. Your study plan is all about helping you be more productive. If you find that your study plan is not as effective as you want, don't get discouraged. It's okay to make changes as you figure out what works best for you.

STEP 2: Choose Your Study Resources

There are numerous textbooks and online resources available for the CBEST Math test, and it may not be clear where to begin. Don't worry! This study guide provides everything you need to fully prepare for your CBEST Math test. In addition to the book content, you can also use Effortless Math's online resources. (video lessons, worksheets, formulas, etc.)

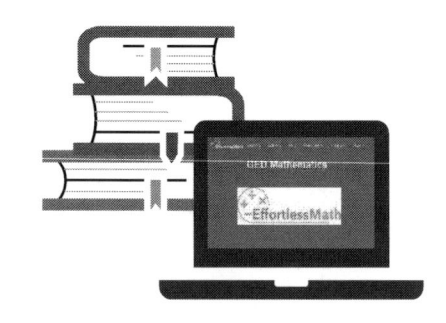

Simply visit EffortlessMath.com/CBEST to find your online CBEST Math resources.

STEP 3: Review, Learn, Practice

This CBEST Math study guide breaks down each subject into specific skills or content areas. For instance, the percent concept is separated into different topics–percent calculation, percent increase and decrease, percent problems, etc. Use this book to help you go over all key math concepts and topics on the CBEST Math test.

As you read each chapter, take notes or highlight the concepts you would like to go over again in the future. If you're unfamiliar with a topic or something is difficult for you, do additional research on it. For each math topic, plenty of instructions, step-by-step guides, and examples are provided to ensure you get a good grasp of the material. You can also find video lessons on the Effortless Math website for each CBEST Math concept.

Quickly review the topics you do understand to get a brush-up of the material. Be sure to do the practice questions provided at the end of every chapter to measure your understanding of the concepts.

STEP 4: Learn and Practice Test-taking Strategies

In the following sections, you will find important test-taking strategies and tips that can help you earn extra points. You'll learn how to think strategically and when to guess if you don't know the answer to a question. Using CBEST Math test-taking strategies and tips can help you raise your score and do well on the test. Apply test taking strategies on the practice tests to help you boost your confidence.

STEP 5: Learn the CBEST Test Format and Take Practice Tests

The *CBEST Test Review* section provides information about the structure of the CBEST test. Read this section to learn more about the CBEST test structure, different test sections, the number of questions in each section, and the section time limits. When you have a prior understanding of the test format and different types of CBEST Math questions, you'll feel more confident when you take the actual exam.

Once you have read through the instructions and lessons and feel like you are ready to go – take advantage of both of the full-length CBEST Math practice tests available in this study guide. Use the practice tests to sharpen your skills and build confidence.

The CBEST Math practice tests offered at the end of the book are formatted similarly to the actual CBEST Math test. When you take each practice test, try to simulate actual testing conditions. To take the practice tests, sit in a quiet space, time yourself, and work through as many of the questions as time allows. The practice tests are followed by detailed answer explanations to help you find your weak areas, learn from your mistakes, and raise your CBEST Math score.

STEP 6: Analyze Your Performance

After taking the practice tests, look over the answer keys and explanations to learn which questions you answered correctly and which you did not. Never be discouraged if you make a few mistakes. See them as a learning opportunity. This will highlight your strengths and weaknesses.

You can use the results to determine if you need additional practice or if you are ready to take the actual CBEST Math test.

EffortlessMath.com

Looking for more?

Visit EffortlessMath.com/CBEST to find hundreds of CBEST Math worksheets, video tutorials, practice tests, CBEST Math formulas, and much more.

Or scan this QR code.

No Registration Required.

CBEST Test Review

The California Basic Educational Skills Test (CBEST) is a computer-based test for Educators who want to gain credentials and teach at public schools. In essence, it is a broad and quick assessment of test takers' academic abilities.

The exam was designed and is administered by the California legislation. The CBEST test contains three sections.

- Math
- Reading
- Writing

CBEST does not test an individual's teaching abilities; it only measures reading skills (comprehension, analysis, and research skills), mathematics skills (calculations and problem solving, etc.), and writing skills that are vital in the education field, either at the elementary, secondary, or higher education levels.

The CBEST Math is comprised of 50 multiple choice questions and test takers have 4 hours to complete all three sections. You may take 1, 2, or all 3 sections in a single test session; you do not have to pass all 3 sections at a single administration.

Use of calculators is prohibited on this examination.

CBEST Mathematics cover the following topics:

- Estimation, measurement, and statistical principles (30%)
- Computation and problem solving (35%)
- Numerical and graphic relationships (35%)

Contents

DAY 1 — Fractions and Mixed Numbers — 1

Simplifying Fractions 2
Adding and Subtracting Fractions... 3
Multiplying and Dividing Fractions 4
Adding Mixed Numbers 5
Subtracting Mixed Numbers 6
Multiplying Mixed Numbers 7
Dividing Mixed Numbers 8
Day 1: Practices 9
Day 1: Answers 12

DAY 2 — Decimals and Integers — 15

Comparing Decimals 16
Rounding Decimals... 17
Adding and Subtracting Decimals 18
Multiplying and Dividing Decimals 19
Adding and Subtracting Integers 20
Multiplying and Dividing Integers... 21
Order of Operations 22
Integers and Absolute Value 23
Day 2: Practices 24
Day 2: Answers 26

DAY 3 — Ratios, Proportions and Percent — 29

Simplifying Ratios 30
Proportional Ratios... 31
Similarity and Ratios 32
Percent Problems 33
Percent of Increase and Decrease... 34
Discount, Tax and Tip... 35
Simple Interest... 36
Day 3: Practices 37
Day 3: Answers 40

DAY 4: Exponents and Variables — 43

- Multiplication Property of Exponents 44
- Division Property of Exponents 45
- Powers of Products and Quotients 46
- Zero and Negative Exponents 47
- Negative Exponents and Negative Bases 48
- Scientific Notation 49
- Radicals 50
- Day 4: Practices 51
- Day 4: Answers 54

DAY 5: Expressions and Variables — 57

- Simplifying Variable Expressions 58
- Simplifying Polynomial Expressions 59
- The Distributive Property 60
- Evaluating One Variable 61
- Evaluating Two Variables 62
- Day 5: Practices 63
- Day 5: Answers 66

DAY 6: Equations and Inequalities — 69

- One–Step Equations 70
- Multi–Step Equations 71
- System of Equations 72
- Graphing Single–Variable Inequalities 73
- One–Step Inequalities 74
- Multi–Step Inequalities 75
- Day 6: Practices 76
- Day 6: Answers 78

DAY 7: Lines and Slope — 81

- Finding Slope 82
- Graphing Lines Using Slope–Intercept Form 83
- Writing Linear Equations 84
- Finding Midpoint 85
- Finding Distance of Two Points 86
- Graphing Linear Inequalities 87
- Day 7: Practices 88
- Day 7: Answers 90

DAY 8: Polynomials — 93

- Simplifying Polynomials 94
- Adding and Subtracting Polynomials 95
- Multiplying Monomials 96
- Multiplying and Dividing Monomials 97
- Multiplying a Polynomial and a Monomial 98
- Multiplying Binomials 99
- Factoring Trinomials 100
- Day 8: Practices 101
- Day 8: Answers 103

DAY 9: Geometry and Solid Figures — 107

- The Pythagorean Theorem 108
- Complementary and Supplementary angles 109
- Parallel lines and Transversals 110
- Triangles 111
- Special Right Triangles 112
- Polygons 113
- Circles 114
- Trapezoids 115
- Cubes 116
- Rectangular Prisms 117
- Cylinder 118
- Day 9: Practices 119
- Day 9: Answers 122

DAY 10: Statistics and Functions — 125

- Mean, Median, Mode, and Range of the Given Data 126
- Pie Graph 127
- Probability Problems 128
- Permutations and Combinations 129
- Function Notation and Evaluation 130
- Adding and Subtracting Functions 131
- Multiplying and Dividing Functions 132
- Composition of Functions 133
- Day 10: Practices 134
- Day 10: Answers 136

Cracking CBEST Math Test -- 139
CBEST Math Practice Test 1 -- 149
CBEST Math Practice Test 2 -- 163
CBEST Math Practice Tests Answer Keys ------------------------------------- 177
CBEST Math Practice Tests Answers and Explanations --------------------- 179

DAY 1: Fractions and Mixed Numbers

Math topics that you'll learn in this chapter:

1. Simplifying Fractions
2. Adding and Subtracting Fractions
3. Multiplying and Dividing Fractions
4. Adding Mixed Numbers
5. Subtracting Mixed Numbers
6. Multiplying Mixed Numbers
7. Dividing Mixed Numbers

Simplifying Fractions

☆ A fraction contains two numbers separated by a bar between them. The bottom number, called the denominator, is the total number of equally divided portions in one whole. The top number, called the numerator, is how many portions you have. And the bar represents the operation of division.

☆ Simplifying a fraction means reducing it to the lowest terms. To simplify a fraction, evenly divide both the top and bottom of the fraction by 2, 3, 5, 7, etc.

☆ Continue until you can't go any further.

Examples:

Example 1. Simplify $\frac{16}{24}$

Solution: To simplify $\frac{16}{24}$, find a number that both 16 and 24 are divisible by. Both are divisible by 8. Then: $\frac{16}{24} = \frac{16 \div 8}{24 \div 8} = \frac{2}{3}$

Example 2. Simplify $\frac{36}{96}$

Solution: To simplify $\frac{36}{96}$, find a number that both 36 and 96 are divisible by. Both are divisible by 6 and 12. Then: $\frac{36}{96} = \frac{36 \div 6}{96 \div 6} = \frac{6}{16}$, 6 and 16 are divisible by 2, then: $\frac{6}{16} = \frac{3}{8}$ or $\frac{36}{96} = \frac{36 \div 12}{96 \div 12} = \frac{3}{8}$

Example 3. Simplify $\frac{43}{129}$

Solution: To simplify $\frac{43}{129}$, find a number that both 43 and 129 are divisible by. Both are divisible by 43, then: $\frac{43}{129} = \frac{43 \div 43}{129 \div 43} = \frac{1}{3}$

bit.ly/3nOGNko

Find more at

EffortlessMath.com

Adding and Subtracting Fractions

★ For "like" fractions (fractions with the same denominator), add or subtract the numerators (top numbers) and write the answer over the common denominator (bottom numbers).

★ Adding and Subtracting fractions with the same denominator:
$$\frac{a}{b} + \frac{c}{b} = \frac{a+c}{b}, \frac{a}{b} - \frac{c}{b} = \frac{a-c}{b}$$

★ Find equivalent fractions with the same denominator before you can add or subtract fractions with different denominators.

★ Adding and Subtracting fractions with different denominators:
$$\frac{a}{b} + \frac{c}{d} = \frac{ad+bc}{bd}, \frac{a}{b} - \frac{c}{d} = \frac{ad-bc}{bd}$$

Examples:

Example 1. Find the sum. $\frac{3}{4} + \frac{2}{3} =$

Solution: These two fractions are "unlike" fractions. (they have different denominators). Use this formula: $\frac{a}{b} + \frac{c}{d} = \frac{ad+cb}{bd}$
Then: $\frac{3}{4} + \frac{2}{3} = \frac{(3)(3)+(4)(2)}{4 \times 3} = \frac{9+8}{12} = \frac{17}{12}$

Example 2. Find the difference. $\frac{4}{7} - \frac{2}{5} =$

Solution: For "unlike" fractions, find equivalent fractions with the same denominator before you can add or subtract fractions with different denominators. Use this formula: $\frac{a}{b} - \frac{c}{d} = \frac{ad-bc}{bd}$
$\frac{4}{7} - \frac{2}{5} = \frac{(4)(5)-(2)(7)}{7 \times 5} = \frac{20-14}{35} = \frac{6}{35}$

Multiplying and Dividing Fractions

✯ **Multiplying fractions:** multiply the top numbers and multiply the bottom numbers. Simplify if necessary. $\frac{a}{b} \times \frac{c}{d} = \frac{a \times c}{b \times d}$

✯ **Dividing fractions:** Keep, Change, Flip

✯ Keep the first fraction, change the division sign to multiplication, and flip the numerator and denominator of the second fraction. Then, solve!

$$\frac{a}{b} \div \frac{c}{d} = \frac{a}{b} \times \frac{d}{c} = \frac{a \times d}{b \times c}$$

Examples:

Example 1. Multiply. $\frac{3}{4} \times \frac{2}{5} =$

Solution: Multiply the top numbers and multiply the bottom numbers.
$\frac{3}{4} \times \frac{2}{5} = \frac{3 \times 2}{4 \times 5} = \frac{6}{20}$, now, simplify: $\frac{6}{20} = \frac{6 \div 2}{20 \div 2} = \frac{3}{10}$

Example 2. Solve. $\frac{2}{3} \div \frac{3}{7} =$

Solution: Keep the first fraction, change the division sign to multiplication, and flip the numerator and denominator of the second fraction.
Then: $\frac{2}{3} \div \frac{3}{7} = \frac{2}{3} \times \frac{7}{3} = \frac{2 \times 7}{3 \times 3} = \frac{14}{9}$

Example 3. Calculate. $\frac{6}{5} \times \frac{2}{3} =$

Solution: Multiply the top numbers and multiply the bottom numbers.
$\frac{6}{5} \times \frac{2}{3} = \frac{6 \times 2}{5 \times 3} = \frac{12}{15}$, simplify: $\frac{12}{15} = \frac{12 \div 3}{15 \div 3} = \frac{4}{5}$

Example 4. Solve. $\frac{4}{5} \div \frac{3}{8} =$

Solution: Keep the first fraction, change the division sign to multiplication, and flip the numerator and denominator of the second fraction.
Then: $\frac{4}{5} \div \frac{3}{8} = \frac{4}{5} \times \frac{8}{3} = \frac{4 \times 8}{5 \times 3} = \frac{32}{15}$

Adding Mixed Numbers

Use the following steps for adding mixed numbers:

☆ Add whole numbers of the mixed numbers.

☆ Add the fractions of the mixed numbers.

☆ Find the Least Common Denominator (LCD) if necessary.

☆ Add whole numbers and fractions.

☆ Write your answer in lowest terms.

Examples:

Example 1. Add mixed numbers. $3\frac{2}{3} + 1\frac{2}{5} =$

Solution: Let's rewriting our equation with parts separated, $3\frac{2}{3} + 1\frac{2}{5} = 3 + \frac{2}{3} + 1 + \frac{2}{5}$. Now, add whole number parts: $3 + 1 = 4$. Add the fraction parts $\frac{2}{3} + \frac{2}{5}$. Rewrite to solve with the equivalent fractions. $\frac{2}{3} + \frac{2}{5} = \frac{10}{15} + \frac{6}{15} = \frac{16}{15}$. The answer is an improper fraction (numerator is bigger than denominator). Convert the improper fraction into a mixed number: $\frac{16}{15} = 1\frac{1}{15}$. Now, combine the whole and fraction parts: $4 + 1\frac{1}{15} = 5\frac{1}{15}$.

Example 2. Find the sum. $2\frac{1}{2} + 1\frac{3}{5} =$

Solution: Rewriting our equation with parts separated, $2 + \frac{1}{2} + 1 + \frac{3}{5}$. Add the whole number parts: $2 + 1 = 3$. Add the fraction parts: $\frac{1}{2} + \frac{3}{5} = \frac{5}{10} + \frac{6}{10} = \frac{11}{10}$. Convert the improper fraction into a mixed number: $\frac{11}{10} = 1\frac{1}{10}$. Now, combine the whole and fraction parts: $3 + 1\frac{1}{10} = 4\frac{1}{10}$.

EffortlessMath.com

Subtracting Mixed Numbers

Use these steps for subtracting mixed numbers.

☆ Convert mixed numbers into improper fractions. $a\frac{c}{b} = \frac{ab+c}{b}$

☆ Find equivalent fractions with the same denominator for unlike fractions. (fractions with different denominators)

☆ Subtract the second fraction from the first one. $\frac{a}{b} - \frac{c}{d} = \frac{ad-bc}{bd}$

☆ Write your answer in lowest terms.

☆ If the answer is an improper fraction, convert it into a mixed number.

Examples:

Example 1. Subtract. $2\frac{1}{5} - 1\frac{2}{3} =$

Solution: Convert mixed numbers into fractions:

$2\frac{1}{5} = \frac{2 \times 5 + 1}{5} = \frac{11}{5}$ and $1\frac{2}{3} = \frac{1 \times 3 + 2}{3} = \frac{5}{3}$

These two fractions are "unlike" fractions. (they have different denominators).

Find equivalent fractions with the same denominator. Use this formula:

$\frac{a}{b} - \frac{c}{d} = \frac{ad-bc}{bd}$

$\frac{11}{5} - \frac{5}{3} = \frac{(11)(3)-(5)(5)}{5 \times 3} = \frac{33-25}{15} = \frac{8}{15}$

Example 2. Find the difference. $2\frac{3}{7} - 1\frac{4}{5} =$

Solution: Convert mixed numbers into fractions:

$2\frac{3}{7} = \frac{2 \times 7 + 3}{7} = \frac{17}{7}$ and $1\frac{4}{5} = \frac{1 \times 5 + 4}{5} = \frac{9}{5}$

Then: $2\frac{3}{7} - 1\frac{4}{5} = \frac{17}{7} - \frac{9}{5} = \frac{(17)(5)-(9)(7)}{7 \times 5} = \frac{85-63}{35} = \frac{22}{35}$

Multiplying Mixed Numbers

Use the following steps for multiplying mixed numbers:

☆ Convert the mixed numbers into fractions. $a\frac{c}{b} = a + \frac{c}{b} = \frac{ab+c}{b}$

☆ Multiply fractions. $\frac{a}{b} \times \frac{c}{d} = \frac{a \times c}{b \times d}$

☆ Write your answer in lowest terms.

☆ If the answer is an improper fraction (numerator is bigger than denominator), convert it into a mixed number.

Examples:

Example 1. Multiply. $3\frac{1}{3} \times 2\frac{3}{5} =$

Solution: Convert mixed numbers into fractions,
$3\frac{1}{3} = \frac{3 \times 3 + 1}{3} = \frac{10}{3}$ and $2\frac{3}{5} = \frac{2 \times 5 + 3}{5} = \frac{13}{5}$. Apply the fractions rule for multiplication:
$\frac{10}{3} \times \frac{13}{5} = \frac{10 \times 13}{3 \times 5} = \frac{130}{15} = \frac{130 \div 5}{15 \div 5} = \frac{26}{3}$
The answer is an improper fraction. Convert it into a mixed number. $\frac{26}{3} = 8\frac{2}{3}$

Example 2. Multiply. $2\frac{3}{4} \times 4\frac{3}{7} =$

Solution: Converting mixed numbers into fractions, $2\frac{3}{4} \times 4\frac{3}{7} = \frac{11}{4} \times \frac{31}{7}$
Apply the fractions rule for multiplication: $\frac{11}{4} \times \frac{31}{7} = \frac{11 \times 31}{4 \times 7} = \frac{341}{28} = 12\frac{5}{28}$

Example 3. Find the product. $4\frac{3}{5} \times 3\frac{4}{6} =$

Solution: Convert mixed numbers to fractions: $4\frac{3}{5} = \frac{23}{5}$ and $3\frac{4}{6} = \frac{22}{6} = \frac{22 \div 2}{6 \div 2} = \frac{11}{3}$.

Multiply two fractions:
$\frac{23}{5} \times \frac{11}{3} = \frac{23 \times 11}{5 \times 3} = \frac{253}{15} = 16\frac{13}{15}$

Dividing Mixed Numbers

Use the following steps for dividing mixed numbers:

☆ Convert the mixed numbers into fractions. $a\frac{c}{b} = a + \frac{c}{b} = \frac{ab+c}{b}$

☆ Divide fractions: Keep, Change, Flip: Keep the first fraction, change the division sign to multiplication, and flip the numerator and denominator of the second fraction. Then, solve! $\frac{a}{b} \div \frac{c}{d} = \frac{a}{b} \times \frac{d}{c} = \frac{a \times d}{b \times c}$

☆ Write your answer in lowest terms.

☆ If the answer is an improper fraction (numerator is bigger than denominator), convert it into a mixed number.

Examples:

Example 1. Solve. $2\frac{2}{3} \div 1\frac{1}{2} =$

Solution: Convert mixed numbers into fractions:
$2\frac{2}{3} = \frac{2 \times 3 + 2}{3} = \frac{8}{3}$ and $1\frac{1}{2} = \frac{1 \times 2 + 1}{2} = \frac{3}{2}$
Keep, Change, Flip: $\frac{8}{3} \div \frac{3}{2} = \frac{8}{3} \times \frac{2}{3} = \frac{8 \times 2}{3 \times 3} = \frac{16}{9}$. The answer is an improper fraction. Convert it into a mixed number: $\frac{16}{9} = 1\frac{7}{9}$

Example 2. Solve. $4\frac{2}{3} \div 1\frac{3}{5} =$

Solution: Convert mixed numbers to fractions, then solve:
$4\frac{2}{3} \div 1\frac{3}{5} = \frac{14}{3} \div \frac{8}{5} = \frac{14}{3} \times \frac{5}{8} = \frac{70}{24} = 2\frac{11}{12}$

Example 3. Solve. $3\frac{2}{5} \div 2\frac{1}{3} =$

Solution: Converting mixed numbers to fractions: $3\frac{2}{5} \div 2\frac{1}{3} = \frac{17}{5} \div \frac{7}{3}$
Keep, Change, Flip: $\frac{17}{5} \div \frac{7}{3} = \frac{17}{5} \times \frac{3}{7} = \frac{17 \times 3}{5 \times 7} = \frac{51}{35} = 1\frac{16}{35}$

bit.ly/2KLPk9k

EffortlessMath.com

Day 1: Practices

✎ Simplify each fraction.

1) $\frac{2}{8} =$

2) $\frac{5}{15} =$

3) $\frac{12}{36} =$

4) $\frac{65}{120} =$

✎ Find the sum or difference.

5) $\frac{3}{10} + \frac{2}{10} =$

6) $\frac{4}{5} + \frac{1}{10} =$

7) $\frac{3}{4} + \frac{6}{20} =$

8) $\frac{4}{9} - \frac{1}{9} =$

9) $\frac{3}{8} - \frac{1}{6} =$

10) $\frac{9}{21} - \frac{2}{7} =$

✎ Find the products or quotients.

11) $\frac{3}{4} \div \frac{9}{12} =$

12) $\frac{7}{10} \div \frac{21}{20} =$

13) $\frac{12}{21} \div \frac{3}{7} =$

14) $\frac{12}{5} \times \frac{10}{24} =$

15) $\frac{33}{36} \times \frac{3}{4} =$

16) $\frac{7}{9} \times \frac{1}{3} =$

✎ Find the sum.

17) $3\frac{1}{2} + 1\frac{3}{4} =$

18) $4\frac{1}{8} + 2\frac{7}{8} =$

19) $4\frac{1}{2} + 2\frac{3}{8} =$

20) $1\frac{2}{21} + 4\frac{4}{7} =$

21) $6\frac{3}{5} + 1\frac{2}{3} =$

22) $2\frac{3}{11} + 3\frac{1}{2} =$

✎ Find the difference.

23) $6\frac{2}{3} - 4\frac{1}{3} =$

24) $5\frac{2}{5} - 3\frac{1}{5} =$

25) $8\frac{1}{2} - 3\frac{1}{4} =$

26) $7\frac{2}{3} - 2\frac{1}{6} =$

✎ **Find the products.**

27) $1\frac{2}{3} \times 2\frac{3}{4} =$

28) $1\frac{1}{6} \times 1\frac{3}{5} =$

29) $4\frac{1}{2} \times 1\frac{2}{3} =$

30) $2\frac{1}{2} \times 4\frac{4}{5} =$

31) $2\frac{1}{5} \times 4\frac{1}{2} =$

32) $1\frac{1}{9} \times 2\frac{3}{5} =$

✎ **Solve.**

33) $3\frac{1}{3} \div 1\frac{2}{3} =$

34) $4\frac{2}{3} \div 2\frac{1}{2} =$

35) $6\frac{1}{5} \div 2\frac{1}{3} =$

36) $2\frac{2}{3} \div 1\frac{4}{9} =$

37) $4\frac{1}{6} \div 2\frac{1}{8} =$

38) $3\frac{2}{5} \div 1\frac{5}{4} =$

39) A pizza cut into 6 parts. David and Sara ordered two pizzas. David ate $\frac{1}{3}$ of his pizza and Sara ate $\frac{1}{2}$ of her pizza. What part of the two pizzas was left?

40) Jake is preparing to run a marathon. He runs $9\frac{1}{3}$ miles on Saturday and two times that many on Monday and Wednesday. Jake wants to run a total of 50 miles this week. How many more miles does he need to run?

41) Last week 21,000 fans attended a football match. This week four times as many bought tickets, but one third of them cancelled their tickets. How many are attending this week?

42) In a bag of small balls $\frac{1}{2}$ are black, $\frac{1}{4}$ are white, $\frac{1}{8}$ are red and the remaining 16 blue. How many balls are white?

Effortless Math Education

Day 1: Answers

1) $\frac{2}{8} = \frac{2 \div 2}{8 \div 2} = \frac{1}{4}$

2) $\frac{5}{15} = \frac{5 \div 5}{15 \div 5} = \frac{1}{3}$

3) $\frac{12}{36} = \frac{12 \div 12}{36 \div 12} = \frac{1}{3}$

4) $\frac{65}{120} = \frac{65 \div 5}{120 \div 5} = \frac{13}{24}$

5) $\frac{3}{10} + \frac{2}{10} = \frac{3+2}{10} = \frac{5}{10} = \frac{5 \div 5}{10 \div 5} = \frac{1}{2}$

6) $\frac{4}{5} + \frac{1}{10} = \frac{4 \times 2}{5 \times 2} + \frac{1}{10} = \frac{8}{10} + \frac{1}{10} = \frac{8+1}{10} = \frac{9}{10}$

7) $\frac{3}{4} + \frac{6}{20} = \frac{3 \times 5}{4 \times 5} + \frac{6}{20} = \frac{15}{20} + \frac{6}{20} = \frac{21}{20}$

8) $\frac{4}{9} - \frac{1}{9} = \frac{4-1}{9} = \frac{3}{9} = \frac{3 \div 3}{9 \div 3} = \frac{1}{3}$

9) $\frac{3}{8} - \frac{1}{6} = \frac{3 \times 6}{8 \times 6} - \frac{1 \times 8}{6 \times 8} = \frac{18}{48} - \frac{8}{48} = \frac{18-8}{48} = \frac{10}{48} = \frac{10 \div 2}{48 \div 2} = \frac{5}{24}$

10) $\frac{9}{21} - \frac{2}{7} = \frac{9}{21} - \frac{2 \times 3}{7 \times 3} = \frac{9}{21} - \frac{6}{21} = \frac{9-6}{21} = \frac{3}{21} = \frac{3 \div 3}{21 \div 3} = \frac{1}{7}$

11) $\frac{3}{4} \div \frac{9}{12} = \frac{3}{4} \times \frac{12}{9} = \frac{36}{36} = 1$

12) $\frac{7}{10} \div \frac{21}{20} = \frac{7}{10} \times \frac{20}{21} = \frac{140}{210} = \frac{140 \div 70}{210 \div 70} = \frac{2}{3}$

13) $\frac{12}{21} \div \frac{3}{7} = \frac{12 \div 3}{21 \div 3} \times \frac{7}{3} = \frac{4}{7} \times \frac{7}{3} = \frac{28}{21} = \frac{28 \div 7}{21 \div 7} = \frac{4}{3}$

14) $\frac{12}{5} \times \frac{10}{24} = \frac{120}{120} = 1$

15) $\frac{33}{36} \times \frac{3}{4} = \frac{33 \div 3}{36 \div 3} \times \frac{3}{4} = \frac{11}{12} \times \frac{3}{4} = \frac{33 \div 3}{48 \div 3} = \frac{11}{16}$

16) $\frac{7}{9} \times \frac{1}{3} = \frac{7}{27}$

17) $3\frac{1}{2} + 1\frac{3}{4} \rightarrow 3 + \frac{1}{2} + 1 + \frac{3}{4} \rightarrow 3 + 1 = 4, \; \frac{1}{2} + \frac{3}{4} = \frac{1 \times 2}{2 \times 2} + \frac{3}{4} = \frac{2}{4} + \frac{3}{4} = \frac{2+3}{4} = \rightarrow$
$\frac{5}{4} = 1\frac{1}{4}, \; 4 + 1\frac{1}{4} = 5\frac{1}{4}$

18) $4\frac{1}{8} + 2\frac{7}{8} \rightarrow 4 + \frac{1}{8} + 2 + \frac{7}{8} \rightarrow 4 + 2 = 6, \; \frac{1}{8} + \frac{7}{8} = \frac{1+7}{8} = \frac{8}{8} = 1 \rightarrow 6 + 1 = 7$

19) $4\frac{1}{2} + 2\frac{3}{8} \rightarrow 4 + \frac{1}{2} + 2 + \frac{3}{8} \rightarrow 4 + 2 = 6, \; \frac{1}{2} + \frac{3}{8} = \frac{1 \times 4}{2 \times 4} + \frac{3}{8} = \frac{4}{8} + \frac{3}{8} = \frac{4+3}{8} = \frac{7}{8} \rightarrow$
$6 + \frac{7}{8} = 6\frac{7}{8}$

20) $1\frac{2}{21} + 4\frac{4}{7} \rightarrow 1 + \frac{2}{21} + 4 + \frac{4}{7} \rightarrow 1 + 4 = 5, \; \frac{2}{21} + \frac{4}{7} = \frac{2}{21} + \frac{4 \times 3}{7 \times 3} = \frac{2}{21} + \frac{12}{21} = \frac{2+12}{21} = \frac{14}{21} =$
$\frac{14 \div 7}{21 \div 7} = \frac{2}{3}$
$\rightarrow 5 + \frac{2}{3} = 5\frac{2}{3}$

Effortless Math Education

21) $6\frac{3}{5} + 1\frac{2}{3} \to 6 + \frac{3}{5} + 1 + \frac{2}{3} \to 6 + 1 = 7, \frac{3}{5} + \frac{2}{3} = \frac{3\times3}{5\times3} + \frac{2\times5}{3\times5} = \frac{9}{15} + \frac{10}{15} = \frac{9+10}{15} = \frac{19}{15} = 1\frac{4}{15} \to$
$7 + 1\frac{4}{15} = 8\frac{4}{15}$

22) $2\frac{3}{11} + 3\frac{1}{2} \to 2 + \frac{3}{11} + 3 + \frac{1}{2} \to 2 + 3 = 5, \frac{3}{11} + \frac{1}{2} = \frac{3\times2}{11\times2} + \frac{1\times11}{2\times11} = \frac{6}{22} + \frac{11}{22} = \frac{6+11}{22} = \frac{17}{22}$
$\to 5 + \frac{17}{22} = 5\frac{17}{22}$

23) $6\frac{2}{3} - 4\frac{1}{3} \to 6 + \frac{2}{3} - 4 - \frac{1}{3} \to 6 - 4 = 2, \frac{2}{3} - \frac{1}{3} = \to \frac{2-1}{3} = \frac{1}{3} \to 2 + \frac{1}{3} = 2\frac{1}{3}$

24) $5\frac{2}{5} - 3\frac{1}{5} \to 5 + \frac{2}{5} - 3 - \frac{1}{5} \to 5 - 3 = 2, \frac{2}{5} - \frac{1}{5} = \frac{1}{5} \to 2 + \frac{1}{5} = 2\frac{1}{5}$

25) $8\frac{1}{2} - 3\frac{1}{4} \to 8 + \frac{1}{2} - 3 - \frac{1}{4} \to 8 - 3 = 5, \frac{1}{2} - \frac{1}{4} = \frac{1\times2}{2\times2} - \frac{1}{4} = \frac{1}{4} \to 5 + \frac{1}{4} = 5\frac{1}{4}$

26) $7\frac{2}{3} - 2\frac{1}{6} \to 7 + \frac{2}{3} - 2 - \frac{1}{6} \to 7 - 2 = 5, \frac{2}{3} - \frac{1}{6} = \frac{2\times2}{3\times2} - \frac{1}{6} = \frac{3\div3}{6\div3} = \frac{1}{2} \to 5 + \frac{1}{2} = 5\frac{1}{2}$

27) $1\frac{2}{3} \times 2\frac{3}{4} \to 1\frac{2}{3} = \frac{1\times3+2}{3} = \frac{5}{3}, 2\frac{3}{4} = \frac{2\times4+3}{4} = \frac{11}{4} \to \frac{5}{3} \times \frac{11}{4} = \frac{5\times11}{3\times4} = \frac{55}{12} = 4\frac{7}{12}$

28) $1\frac{1}{6} \times 1\frac{3}{5} \to 1\frac{1}{6} = \frac{1\times6+1}{6} = \frac{7}{6}, 1\frac{3}{5} = \frac{1\times5+3}{5} = \frac{8}{5} \to \frac{7}{6} \times \frac{8}{5} = \frac{7\times8}{6\times5} = \frac{56}{30} = \frac{56\div2}{30\div2} = \frac{28}{15} = 1\frac{13}{15}$

29) $4\frac{1}{2} \times 1\frac{2}{3} \to 4\frac{1}{2} = \frac{4\times2+1}{2} = \frac{9}{2}, 1\frac{2}{3} = \frac{1\times3+2}{3} = \frac{5}{3} \to \frac{9}{2} \times \frac{5}{3} = \frac{9\times5}{2\times3} = \frac{45}{6} = \frac{45\div3}{6\div3} = \frac{15}{2} = 7\frac{1}{2}$

30) $2\frac{1}{2} \times 4\frac{4}{5} \to 2\frac{1}{2} = \frac{2\times2+1}{2} = \frac{5}{2}, 4\frac{4}{5} = \frac{4\times5+4}{5} = \frac{24}{5} \to \frac{5}{2} \times \frac{24}{5} = \frac{5\times24}{2\times5} = \frac{120}{10} = 12$

31) $2\frac{1}{5} \times 4\frac{1}{2} \to 2\frac{1}{5} = \frac{2\times5+1}{5} = \frac{11}{5}, 4\frac{1}{2} = \frac{4\times2+1}{2} = \frac{9}{2} \to \frac{11}{5} \times \frac{9}{2} = \frac{11\times9}{5\times2} = \frac{99}{10} = 9\frac{9}{10}$

32) $1\frac{1}{9} \times 2\frac{3}{5} \to 1\frac{1}{9} = \frac{1\times9+1}{9} = \frac{10}{9}, 2\frac{3}{5} = \frac{2\times5+3}{5} = \frac{13}{5} \to \frac{10}{9} \times \frac{13}{5} = \frac{10\times13}{9\times5} = \frac{130}{45} = \frac{130\div5}{45\div5} \to$
$\frac{26}{9} = 2\frac{8}{9}$

33) $3\frac{1}{3} \div 1\frac{2}{3} \to 3\frac{1}{3} = \frac{3\times3+1}{3} = \frac{10}{3}, 1\frac{2}{3} = \frac{1\times3+2}{3} = \frac{5}{3} \to \frac{10}{3} \div \frac{5}{3} = \frac{10}{3} \times \frac{3}{5} = \frac{30}{15} = 2$

34) $4\frac{2}{3} \div 2\frac{1}{2} \to 4\frac{2}{3} = \frac{4\times3+2}{3} = \frac{14}{3}, 2\frac{1}{2} = \frac{2\times2+1}{2} = \frac{5}{2} \to \frac{14}{3} \div \frac{5}{2} = \frac{14}{3} \times \frac{2}{5} = \frac{28}{15} = 1\frac{13}{15}$

35) $6\frac{1}{5} \div 2\frac{1}{3} \to 6\frac{1}{5} = \frac{6\times5+1}{5} = \frac{31}{5}, 2\frac{1}{3} = \frac{2\times3+1}{3} = \frac{7}{3} \to \frac{31}{5} \div \frac{7}{3} = \frac{31}{5} \times \frac{3}{7} = \frac{93}{35} = 2\frac{23}{35}$

36) $2\frac{2}{3} \div 1\frac{4}{9} \to 2\frac{2}{3} = \frac{2\times3+2}{3} = \frac{8}{3}, 1\frac{4}{9} = \frac{1\times9+4}{9} = \frac{13}{9} \to \frac{8}{3} \div \frac{13}{9} = \frac{8}{3} \times \frac{9}{13} = \frac{72\div3}{39\div3} = \frac{24}{13} = 1\frac{11}{13}$

37) $4\frac{1}{6} \div 2\frac{1}{8} \to 4\frac{1}{6} = \frac{4\times6+1}{6} = \frac{25}{6}, 2\frac{1}{8} = \frac{2\times8+1}{8} = \frac{17}{8} \to \frac{25}{6} \div \frac{17}{8} = \frac{25}{6} \times \frac{8}{17} =$
$\frac{200\div2}{102\div2} \to \frac{100}{51} = 1\frac{49}{51}$

38) $3\frac{2}{5} \div 1\frac{5}{4} \to 3\frac{2}{5} = \frac{3\times5+2}{5} = \frac{17}{5}, 1\frac{5}{4} = \frac{1\times4+5}{4} = \frac{9}{4} \to \frac{17}{5} \div \frac{9}{4} = \frac{17}{5} \times \frac{4}{9} = \frac{68}{45} = 1\frac{23}{45}$

39) David ate $\frac{1}{3}$ of 6 parts of his pizza $\to \frac{1}{3} \times 6 = \frac{6\div 3}{3\div 3} = 2 \to$ It means 2 parts out of 6 parts and left 4 parts. Sara ate $\frac{1}{2}$ of 6 parts of her pizza:

$\to \frac{1}{2} \times 6 = \frac{6\div 2}{2\div 2} = 3 \to$ It means 3 parts out of 6 parts and left 3 parts. Therefore, they ate (2 + 3) parts out of (6 + 6) parts of their pizza and left (4 + 3) parts out of (6 + 6) parts of their pizza that equals to: $\frac{7}{12}$

40) Jake run $9\frac{1}{3}$ miles on Saturday and $2 \times \left(9\frac{1}{3}\right)$ miles on Monday and Wednesday. Jake wants to run a total of 50 miles this week. Therefore, $9\frac{1}{3} + 2 \times \left(9\frac{1}{3}\right)$ should be subtracted from 50:

$50 - \left(9\frac{1}{3} + \left(2 \times 9\frac{1}{3}\right)\right) = 50 - \left(\frac{9\times3+1}{3} + \left(2 \times \frac{9\times3+1}{3}\right)\right) = 50 - \left(\frac{28}{3} + \frac{56}{3}\right) = 50 - \left(\frac{28+56}{3}\right) = 50 - \left(\frac{84}{3}\right) = 50 - (28) = 22$ miles.

41) Four times of 21,000 is 84,000. One third of them cancelled their tickets. One third of 84,000 Equals 28,000. ($\frac{1}{3} \times 84,000 = 28,000$).

84,000 − 28,000 = 56,000 fans are attending this week.

42) Let x be the total number of balls. Then: $\frac{1}{2}x + \frac{1}{4}x + \frac{1}{8}x + 16 = x$

$\left(\frac{1}{2} + \frac{1}{4} + \frac{1}{8}\right)x + 16 = x \to \left(\frac{1\times 4}{2\times 4} + \frac{1\times 2}{4\times 2} + \frac{1}{8}\right)x + 16 = x \to$

$\left(\frac{4}{8} + \frac{2}{8} + \frac{1}{8}\right)x + 16 = x \to \left(\frac{7}{8}\right)x + 16 = x \to 16 = x - \frac{7}{8}x \to 16 = \frac{1}{8}x$

$\to$ Multiply both sides by 8: $16 \times 8 = \frac{1}{8}x \times 8 \to 128 = x$

x is the total number of balls. Therefore, number of white balls is:

$\frac{1}{4}x = \frac{1}{4} \times 128 = 32$

DAY 2: Decimals and Integers

Math topics that you'll learn in this chapter:

1. Comparing Decimals
2. Rounding Decimals
3. Adding and Subtracting Decimals
4. Multiplying and Dividing Decimals
5. Adding and subtracting Integers
6. Multiplying and Dividing Integers
7. Order of Operations
8. Integers and Absolute Value

Comparing Decimals

☆ A decimal is a fraction written in a special form. For example, instead of writing $\frac{1}{2}$ you can write: 0.5

☆ A Decimal Number contains a Decimal Point. It separates the whole number part from the fractional part of a decimal number.

☆ Let's review decimal place values: Example: 45.3861

    4: tens   5: ones   3: tenths

    8: hundredths   6: thousandths   1: tens thousandths

☆ To compare two decimals, compare each digit of two decimals in the same place value. Start from left. Compare hundreds, tens, ones, tenth, hundredth, etc.

☆ To compare numbers, use these symbols:

Equal to =   Less than <   Greater than >

Less than or equal ≤   Greater than or equal ≥

Examples:

Example 1. Compare 0.05 and 0.50.

Solution: 0.50 is greater than 0.05, because the tenth place of 0.50 is 5, but the tenth place of 0.05 is zero. Then: 0.05 < 0.50

Example 2. Compare 0.0512 and 0.181.

Solution: 0.181 is greater than 0.0512, because the tenth place of 0.181 is 1, but the tenth place of 0.0512 is zero. Then: 0.0512 < 0.181

EffortlessMath.com

Rounding Decimals

☆ We can round decimals to a certain accuracy or number of decimal places. This is used to make calculations easier to do and results easier to understand when exact values are not too important.

☆ First, you'll need to remember your place values: For example: 12.4869

    1: tens   2: ones   4: tenths

    8: hundredths   6: thousandths   9: tens thousandths

☆ To round a decimal, first find the place value you'll round to.

☆ Find the digit to the right of the place value you're rounding to. If it is 5 or bigger, add 1 to the place value you're rounding to and remove all digits on its right side. If the digit to the right of the place value is less than 5, keep the place value and remove all digits on the right.

Examples:

Example 1. Round 3.2568 to the thousandth-place value.

Solution: First, look at the next place value to the right, (tens thousandths). It's 8 and it is greater than 5. Thus add 1 to the digit in the thousandth place. The thousandth place is 6. → $6 + 1 = 7$, then, the answer is 3.257.

Example 2. Round 2.3628 to the nearest hundredth.

Solution: First, look at the digit to the right of hundredth (thousandths place value). It's 2 and it is less than 5, thus remove all the digits to the right of hundredth place. Then, the answer is 2.36.

Adding and Subtracting Decimals

☆ Line up the decimal numbers.

☆ Add zeros to have the same number of digits for both numbers if necessary.

☆ Remember your place values: For example: 73.5196

    7: tens   3: ones   5: tenths

    1: hundredths   9: thousandths   6: tens thousandths

☆ Add or subtract using column addition or subtraction.

Examples:

Example 1. Add. $2.6 + 3.25 =$

Solution: First, line up the numbers: $\begin{array}{r} 2.6 \\ +3.25 \\ \hline \end{array}$ →Add a zero to have the same number of digits for both numbers. $\begin{array}{r} 2.60 \\ +3.25 \\ \hline \end{array}$ →Start with the hundredths place: $0 + 5 = 5$, $\begin{array}{r} 2.60 \\ +3.25 \\ \hline 5 \end{array}$ →Continue with tenths place: $6 + 2 = 8$, $\begin{array}{r} 2.60 \\ +3.25 \\ \hline .85 \end{array}$ →Add the ones place: $2 + 3 = 5$, $\begin{array}{r} 2.60 \\ +3.25 \\ \hline 5.85 \end{array}$. The answer is 5.85.

Example 2. Find the difference. $4.26 - 3.12 =$

Solution: First, line up the numbers: $\begin{array}{r} 4.26 \\ -3.12 \\ \hline \end{array}$ →Start with the hundredths place: $6 - 2 = 4$, $\begin{array}{r} 4.26 \\ -3.12 \\ \hline 4 \end{array}$ → Continue with tenths place. $2 - 1 = 1$, $\begin{array}{r} 4.26 \\ -3.12 \\ \hline .14 \end{array}$ →Subtract the ones place. $4 - 3 = 1$, $\begin{array}{r} 4.26 \\ -3.12 \\ \hline 1.14 \end{array}$

Multiplying and Dividing Decimals

For multiplying decimals:

☆ Ignore the decimal point and set up and multiply the numbers as you do with whole numbers.

☆ Count the total number of decimal places in both of the factors.

☆ Place the decimal point in the product.

For dividing decimals:

☆ If the divisor is not a whole number, move the decimal point to the right to make it a whole number. Do the same for the dividend.

☆ Divide similar to whole numbers.

Examples:

Example 1. Find the product. $0.53 \times 0.32 =$

Solution: Set up and multiply the numbers as you do with whole numbers. Line up the numbers: $\begin{array}{r}53\\ \times 32\end{array}$ →Start with the ones place then continue with other digits → $\begin{array}{r}53\\ \times 32\\ \hline 1,696\end{array}$. Count the total number of decimal places in both of the factors. There are four decimals digits. (two for each factor 0.53 and 0.32) Then: $0.53 \times 0.32 = 0.1696$

Example 2. Find the quotient. $1.50 \div 0.5 =$

Solution: The divisor is not a whole number. Multiply it by 10 to get 5:
→ $0.5 \times 10 = 5$
Do the same for the dividend to get 15. → $1.50 \times 10 = 15$
Now, divide $15 \div 5 = 3$. The answer is 3.

Adding and Subtracting Integers

☆ Integers include zero, counting numbers, and the negative of the counting numbers $\{..., -3, -2, -1, 0, 1, 2, 3, ...\}$

☆ Add a positive integer by moving to the right on the number line. (you will get a bigger number)

☆ Add a negative integer by moving to the left on the number line. (you will get a smaller number)

☆ Subtract an integer by adding its opposite.

Number line

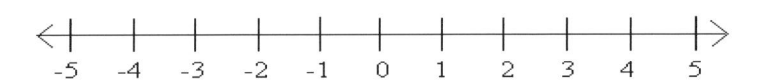

Examples:

Example 1. Solve. $(-3) - (-5) =$

Solution: Keep the first number and convert the sign of the second number to its opposite. (change subtraction into addition. Then: $(-3) + 5 = 2$

Example 2. Solve. $5 + (2 - 8) =$

Solution: First, subtract the numbers in brackets, $2 - 8 = -6$
Then: $5 + (-6) = \rightarrow$ change addition into subtraction: $5 - 6 = -1$

Example 3. Solve. $(8 - 15) + 14 =$

Solution: First, subtract the numbers in brackets, $8 - 15 = -7$
Then: $-7 + 14 = \rightarrow -7 + 14 = 7$

Example 4. Solve. $18 + (-5 - 17) =$

Solution: First, subtract the numbers in brackets, $-5 - 17 = -22$
Then: $18 + (-22) = \rightarrow$ change addition into subtraction: $18 - 22 = -4$

Multiplying and Dividing Integers

Use the following rules for multiplying and dividing integers:

☆ (negative) × (negative) = positive

☆ (negative) ÷ (negative) = positive

☆ (negative) × (positive) = negative

☆ (negative) ÷ (positive) = negative

☆ (positive) × (positive) = positive

☆ (positive) ÷ (negative) = negative

Examples:

Example 1. Solve. $5 \times (-2) =$

Solution: Use this rule: (positive) × (negative) = negative.
Then: $(5) \times (-2) = -10$

Example 2. Solve. $(-2) + (-30 \div 6) =$

Solution: First, divide −30 by 6, the numbers in brackets, use this rule: (negative) ÷ (positive) = negative. Then: $-30 \div 6 = -5$
$(-2) + (-30 \div 6) = (-2) + (-5) = -2 - 5 = -7$

Example 3. Solve. $(13 - 16) \times (-3) =$

Solution: First, subtract the numbers in brackets,
$13 - 16 = -3 \rightarrow (-3) \times (-3) =$
Now use this rule: (negative) × (negative) = positive $\rightarrow (-3) \times (-3) = 9$

Example 4. Solve. $(18 - 3) \div (-5) =$

Solution: First, subtract the numbers in brackets,
$18 - 3 = 15 \rightarrow (15) \div (-5) =$
Now use this rule:
(positive) ÷ (negative) = negative $\rightarrow (15) \div (-5) = -3$

Order of Operations

✯ In Mathematics, "operations" are addition, subtraction, multiplication, division, exponentiation (written as b^n) and grouping.

✯ When there is more than one math operation in an expression, use PEMDAS: (to memorize this rule, remember the phrase "Please Excuse My Dear Aunt Sally".)

- Parentheses
- Exponents
- Multiplication and Division (from left to right)
- Addition and Subtraction (from left to right)

Examples:

Example 1. Calculate. $(3 + 5) \div (8 \div 4) =$

Solution: First, simplify inside parentheses:
$(3 + 5) \div (8 \div 4) = (8) \div (8 \div 4) = (8) \div (2)$. Then: $(8) \div (2) = 4$

Example 2. Solve. $(4 \times 3) - (14 - 3) =$

Solution: First, calculate within parentheses: $(4 \times 3) - (14 - 3) = (12) - (11)$, Then: $(12) - (11) = 1$

Example 3. Calculate. $-3[(6 \times 5) \div (5 \times 3)] =$

Solution: First, calculate within parentheses:
$-3[(6 \times 5) \div (5 \times 3)] = -3[(30) \div (5 \times 3)] = -3[(30) \div (15)] = -3[2]$
Multiply -3 and 2. Then: $-3[2] = -6$

Example 4. Solve. $(32 \div 4) + (-25 + 5) =$

Solution: First, calculate within parentheses:
$(32 \div 4) + (-25 + 5) = (8) + (-20)$. Then: $(8) - (20) = -12$

Integers and Absolute Value

☆ The absolute value of a number is its distance from zero, in either direction, on the number line. For example, the distance of 9 and −9 from zero on number line is 9.

☆ The absolute value of an integer is the numerical value without its sign. (negative or positive)

☆ The vertical bar is used for absolute value as in $|x|$.

☆ The absolute value of a number is never negative; because it only shows, "how far the number is from zero".

Examples:

Example 1. Calculate. $|15 − 3| \times 6 =$

Solution: First, solve $|15 − 3| \rightarrow |15 − 3| = |12|$, the absolute value of 12 is 12, $|12| = 12$. Then: $12 \times 6 = 72$

Example 2. Solve. $|−35| \times |6 − 10| =$

Solution: First, find $|−35| \rightarrow$ the absolute value of −35 is 35. Then: $|−35| = 35$, $|−35| \times |6 − 10| =$
Now, calculate $|6 − 10| \rightarrow |6 − 10| = |−4|$, the absolute value of −4 is 4. $|−4| = 4$
Then: $35 \times 4 = 140$

Example 3. Solve. $|12 − 6| \times \frac{|−4 \times 5|}{3} =$

Solution: First, calculate $|12 − 6| \rightarrow |12 − 6| = |6|$, the absolute value of 6 is 6, $|6| = 6$. Then: $6 \times \frac{|−4 \times 5|}{3} =$
Now calculate $|−4 \times 5| \rightarrow |−4 \times 5| = |−20|$, the absolute value of −20 is 20, $|−20| = 20$. Then: $6 \times \frac{20}{3} = \frac{6 \times 20}{3} = \frac{120}{3} = 40$

Day 2: Practices

✍ Compare. Use >, =, and <

1) 0.3 ☐ 0.2
2) 0.98 ☐ 0.71
3) 5.01 ☐ 5.0100
4) 0.427 ☐ 0.435

✍ Round each decimal to the nearest whole number.

5) 4.9
6) 6.3
7) 75.66
8) 93.03

✍ Find the sum or difference.

9) 2.5 + 11.1 =
10) 12.83 + 14.11 =
11) 13.8 − 9.2 =
12) 43.55 − 21.32 =

✍ Find the product or quotient.

13) 5.1 × 0.2 =
14) 0.35 × 1.2 =
15) 2.1 ÷ 0.3
16) 25.5 ÷ 0.5

✍ Find each sum or difference.

17) −6 + 17 =
18) 12 − 21 =
19) 31 − (−4) =
20) (7 + 5) + (8 − 3) =
21) (2 − 3) − (15 − 11) =
22) (−8 − 7) − (−6 − 2) =

EffortlessMath.com

✎ Solve.

23) $2 \times (-4) =$

24) $(-5) \times (-3) =$

25) $(-15) \div 5 =$

26) $(-4) \times (-5) \times (-2) =$

27) $(-6 + 36) \div (-2) =$

28) $(-25 + 5) \times (-5 - 3) =$

✎ Evaluate each expression.

29) $5 - (2 \times 3) =$

30) $(6 \times 5) - 8 =$

31) $(-5 \times 3) + 4 =$

32) $(-35 \div 5) - (12 + 2) =$

33) $4 \times [(2 \times 3) \div (-3 + 1)] =$

34) $35 \div [(6 - 1) \times (7 - 8)] =$

✎ Find the answers.

35) $|-3| + |7 - 9| =$

36) $|8 - 10| + |6 - 7| =$

37) $|-6 + 10| - |-9 - 3| =$

38) $3 + |2 - 1| + |3 - 12| =$

39) $-8 - |3 - 6| + |2 + 3| =$

40) $|-6| \times |-5.4| =$

41) $|3 \times (-4)| \times \frac{8}{3} =$

42) $|(-2) \times (-2)| \times \frac{1}{4} =$

43) $|-8| + |(-8) \times 2| =$

44) $|(-3) \times (-5)| \times |(-3) \times (-4)| =$

✎ Find the answers.

45) Round 4.2873 to the thousandth-place value

46) $[6 \times (-16) + 8] - (-4) + [4 \times 5] \div 2 =$

Effortless Math Education

Day 2: Answers

1) $0.3 > 0.2$
2) $0.98 > 0.71$
3) $5.01 = 5.0100$
4) $0.427 < 0.435$
5) $4.9 \approx 5$
6) $6.3 \approx 6$
7) $75.66 \approx 76$
8) $93.03 \approx 93$

9) $\begin{array}{r}2.5\\+11.1\\\hline\end{array} \to 5+1=6 \to \begin{array}{r}2.5\\+11.1\\\hline .6\end{array} \to 2+1=3 \to \begin{array}{r}2.5\\+11.1\\\hline 3.6\end{array} \to 0+1=1 \to \begin{array}{r}2.5\\+11.1\\\hline 13.6\end{array}$

10) $\begin{array}{r}12.83\\+14.11\\\hline\end{array} \to 3+1=4 \to \begin{array}{r}12.83\\+14.11\\\hline 4\end{array} \to 8+1=9 \to \begin{array}{r}12.83\\+14.11\\\hline .94\end{array} \to 2+4=6 \to \begin{array}{r}12.83\\+14.11\\\hline 6.94\end{array} \to 1+1=2 \to \begin{array}{r}12.83\\+14.11\\\hline 26.94\end{array}$

11) $\begin{array}{r}13.8\\-9.2\\\hline\end{array} \to 8-2=6 \to \begin{array}{r}13.8\\-9.2\\\hline .6\end{array} \to 13-9=4 \to \begin{array}{r}13.8\\-9.2\\\hline 4.6\end{array}$

12) $\begin{array}{r}43.55\\-21.32\\\hline\end{array} \to 5-2=3 \to \begin{array}{r}43.55\\-21.32\\\hline 3\end{array} \to 5-3=2 \to \begin{array}{r}43.55\\-21.32\\\hline 23\end{array} \to 3-1=2 \to \begin{array}{r}43.55\\-21.32\\\hline 2.23\end{array} \to 4-2=2 \to \begin{array}{r}43.55\\-21.32\\\hline 22.23\end{array}$

13) $5.1 \times 0.2 \to \begin{array}{r}51\\\times 2\\\hline 102\end{array} \to 5.1 \times 0.2 = 1.02$

14) $0.35 \times 1.2 \to \begin{array}{r}35\\\times 12\\\hline 70\\+350\\\hline 420\end{array} \to 0.35 \times 1.2 = 0.420$

15) $2.1 \div 0.3 \to \frac{2.1 \times 10}{0.3 \times 10} = \frac{21}{3} = 7$

16) $25.5 \div 0.5 \to \frac{25.50 \times 10}{0.5 \times 10} = \frac{255}{5} = 51$

17) $-6 + 17 = 17 - 6 = 11$

18) $12 - 21 = -9$

19) $31 - (-4) = 31 + 4 = 35$

20) $(7 + 5) + (8 - 3) = 12 + 5 = 17$

21) $(2 - 3) - (15 - 11) = (-1) - (4) = -1 - 4 = -5$

22) $(-8 - 7) - (-6 - 2) = (-15) - (-8) = -15 + 8 = -7$

23) Use this rule: (positive) × (negative) = negative → $2 \times (-4) = -8$

24) Use this rule: (negative) × (negative) = positive → $(-5) \times (-3) = 15$

25) Use this rule: (negative) ÷ (positive) = negative → $(-15) \div 5 = -3$

26) Use these rules: [(negative) × (negative) = positive] and [(positive) × (negative) = negative]→ $(-4) \times (-5) \times (-2) = (20) \times (-2) = -40$

27) Use this rule: (positive)÷ (negative) = negative→
$(-6 + 36) \div (-2) = (30) \div (-2) = -15$

28) Use this rule: (negative) × (negative) = positive→
$(-25 + 5) \times (-5 - 3) = (-20) \times (-8) = -160$

29) $5 - (2 \times 3) = 5 - 6 = -1$

30) $(6 \times 5) - 8 = 30 - 8 = 22$

31) Use this rule: (negative) × (positive) = negative → $(-5 \times 3) + 4 = -15 + 4 = -11$

32) Use this rule: (negative) ÷ (positive) = negative →
$(-35 \div 5) - (12 + 2) = -7 - 14 = -21$

33) Use these rules: [(positive) × (negative) = negative] and [(positive) ÷ (negative) = negative]→ 4 × [(2 × 3) ÷ (−3 + 1)] = 4 × [6 ÷ (−2)] = 4 × (−3) = −12

34) Use these rules: [(positive) × (negative) = negative] and [(positive) ÷ (negative) = negative]→ 35 ÷ [(6 − 1) × (7 − 8)] = 35 ÷ [5 × (−1)] = 35 ÷ (−5) = −7

35) |−3| + |7 − 9| = 3 + |−2| = 3 + 2 = 5

36) |8 − 10| + |6 − 7| = |−2| + |−1| = 2 + 1 = 3

37) |−6 + 10| − |−9 − 3| = |4| − |−12| = 4 − (12) = 4 − 12 = −8

38) 3 + |2 − 1| + |3 − 12| = 3 + |1| + |−9| = 3 + 1 + (9) = 3 + 1 + 9 = 13

39) −8 − |3 − 6| + |2 + 3| = −8 − |−3| + |5| = −8 − (3) + 5 = −8 − 3 + 5 = −11 + 5 = −6

40) |−6| × |−5.4| = 6 × 5.4 = 32.4

41) |3 × (−4)| × $\frac{8}{3}$ = |−12| × $\frac{8}{3}$ = 12 × $\frac{8}{3}$ = $\frac{12 \times 8}{3}$ = 32

42) |(−2) × (−2)| × $\frac{1}{4}$ = |4| × $\frac{1}{4}$ = 4 × $\frac{1}{4}$ = 1

43) |−8| + |(−8) × 2| = 8 + |−16| = 8 + 16 = 24

44) |(−3) × (−5)| × |(−3) × (−4)| = |15| × |12| = 15 × 12 = 180

45) 4.2873 ≈ 4.287

46) [6 × (−16) + 8] − (−4) + [4 × 5] ÷ 2 = [(−96) + 8] − (−4) + [4 × 5] ÷ 2 = (−88) − (−4) + (20) ÷ 2 = (−88) − (−4) + 10 = (−88) + 4 + (10) = −84 + 10 = −74

DAY 3: Ratios, Proportions and Percent

Math topics that you'll learn in this chapter:

1. Simplifying Ratios
2. Proportional Ratios
3. Similarity and Ratios
4. Percent Problems
5. Percent of increase and Decrease
6. Discount, Tax and Tip
7. Simple Interest

Simplifying Ratios

☆ Ratios are used to make comparisons between two numbers.

☆ Ratios can be written as a fraction, using the word "to", or with a colon. Example: $\frac{3}{4}$ or "3 to 4" or 3:4

☆ You can calculate equivalent ratios by multiplying or dividing both sides of the ratio by the same number.

Examples:

Example 1. Simplify. $10:2 =$

Solution: Both numbers 10 and 2 are divisible by $2 \Rightarrow 10 \div 2 = 5, 2 \div 2 = 1$. Then: $10:2 = 5:1$

Example 2. Simplify. $\frac{6}{33} =$

Solution: Both numbers 6 and 33 are divisible by $3 \Rightarrow 33 \div 3 = 11, 6 \div 3 = 2$. Then: $\frac{6}{33} = \frac{2}{11}$

Example 3. There are 30 students in a class and 12 are girls. Find the ratio of girls to boys in that class.

Solution: Subtract 12 from 30 to find the number of boys in the class.
$30 - 12 = 18$. There are 18 boys in the class. So, the ratio of girls to boys is $12:18$. Now, simplify this ratio. Both 18 and 12 are divisible by 6.
Then: $18 \div 6 = 3$, and $12 \div 6 = 2$. In the simplest form, this ratio is $2:3$

Example 4. A recipe calls for butter and sugar in the ratio $2:3$. If you're using 6 cups of butter, how many cups of sugar should you use?

Solution: Since you use 6 cups of butter, or 3 times as much, you need to multiply the amount of sugar by 3. Then: $3 \times 3 = 9$.
So, you need to use 9 cups of sugar. You can solve this using equivalent fractions: $\frac{2}{3} = \frac{6}{9}$

Proportional Ratios

☆ Two ratios are proportional if they represent the same relationship.

☆ A proportion means that two ratios are equal. It can be written in two ways: $\frac{a}{b} = \frac{c}{d}$ $a:b = c:d$

☆ The proportion $\frac{a}{b} = \frac{c}{d}$ can be written as: $a \times d = c \times b$

Examples:

Example 1. Solve this proportion for x. $\frac{3}{4} = \frac{9}{x}$

Solution: Use cross multiplication: $\frac{3}{4} = \frac{9}{x} \Rightarrow 3 \times x = 4 \times 9 \Rightarrow 3x = 36$
Divide both sides by 3 to find x: $x = \frac{36}{3} \Rightarrow x = 12$

Example 2. If a box contains red and blue balls in ratio of $4:7$ red to blue, how many red balls are there if 49 blue balls are in the box?

Solution: Write a proportion and solve. $\frac{4}{7} = \frac{x}{49}$
Use cross multiplication: $4 \times 49 = 7 \times x \Rightarrow 196 = 7x$
Divide to find x: $x = \frac{196}{7} \Rightarrow x = 28$. There are 28 red balls in the box.

Example 3. Solve this proportion for x. $\frac{5}{8} = \frac{20}{x}$

Solution: Use cross multiplication: $\frac{5}{8} = \frac{20}{x} \Rightarrow 5 \times x = 8 \times 20 \Rightarrow 5x = 160$
Divide to find x: $x = \frac{160}{5} \Rightarrow x = 32$

Example 4. Solve this proportion for x. $\frac{7}{9} = \frac{21}{x}$

Solution: Use cross multiplication: $\frac{7}{9} = \frac{21}{x} \Rightarrow 7 \times x = 9 \times 21 \Rightarrow 7x = 189$
Divide to find x: $x = \frac{189}{7} \Rightarrow x = 27$

Similarity and Ratios

✯ Two figures are similar if they have the same shape.

✯ Two or more figures are similar if the corresponding angles are equal, and the corresponding sides are in proportion.

Examples:

Example 1. The following triangles are similar. What is the value of the unknown side?

Solution: Find the corresponding sides and write a proportion.
$\frac{9}{18} = \frac{8}{x}$. Now, use the cross product to solve for x:
$\frac{9}{18} = \frac{8}{x} \to 9 \times x = 18 \times 8 \to 9x = 144$. Divide both sides by 9. Then: $9x = 144 \to x = \frac{144}{9} \to x = 16$

The missing side is 16.

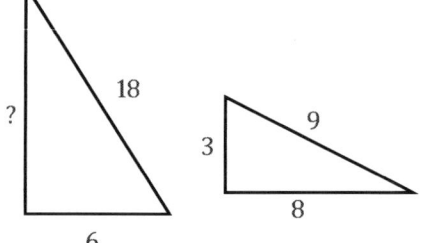

Example 2. Two rectangles are similar. The first is 4 feet wide and 12 feet long. The second is 8 feet wide. What is the length of the second rectangle?

Solution: Let's put x for the length of the second rectangle. Since two rectangles are similar, their corresponding sides are in proportion. Write a proportion and solve for the missing number.
$\frac{4}{8} = \frac{12}{x} \to 4x = 8 \times 12 \to 4x = 96 \to x = \frac{96}{4} = 24$

The length of the second rectangle is 24 feet.

Percent Problems

✯ Percent is a ratio of a number and 100. It always has the same denominator, 100. The percent symbol is "%".

✯ Percent means "per 100". So, 20% is $\frac{20}{100}$.

✯ In each percent problem, we are looking for the base, or the part or the percent.

✯ Use these equations to find each missing section in a percent problem:

- Base = Part ÷ Percent
- Part = Percent × Base
- Percent = Part ÷ Base

Examples:

Example 1. What is 30% of 60?

Solution: In this problem, we have the percent (30%) and the base (60) and we are looking for the "part". Use this formula: $Part = Percent \times Base$.
Then: $Part = 30\% \times 60 = \frac{30}{100} \times 60 = 0.30 \times 60 = 18$. The answer: 30% of 60 is 18.

Example 2. 20 is what percent of 400?

Solution: In this problem, we are looking for the percent. Use this equation:
$Percent = Part \div Base \rightarrow Percent = 20 \div 400 = 0.05 = 5\%$.
Then: 20 is 5 percent of 400.

Example 3. 70 is 25 percent of what number?

Solution: In this problem, we are looking for the base. Use this equation:
$Base = Part \div Percent \rightarrow Base = 70 \div 25\% = 70 \div 0.25 = 280$
Then: 70 is 25 percent of 280.

Percent of Increase and Decrease

☆ Percent of change (increase or decrease) is a mathematical concept that represents the degree of change over time.

☆ To find the percentage of increase or decrease:
  1. New Number−Original Number
  2. (The result÷Original Number)× 100

☆ Or use this formula: Percent of change=$\frac{new\ number-original\ number}{original\ number}\times 100$

☆ Note: If your answer is a negative number, then this is a percentage decrease. If it is positive, then this is a percentage increase.

Examples:

Example 1. The price of a shirt increases from $40 to $44. What is the percentage increase?

Solution: First, find the difference: 44 − 40 = 4

Then: (4 ÷ 40) × 100 = $\frac{4}{40}$ × 100 = 10. The percentage increase is 10%. It means that the price of the shirt increased by 10%.

Example 2. The price of a table decreased from $50 to $25. What is the percent of decrease?

Solution: Use this formula:

$Percent\ of\ change = \frac{new\ number - original\ number}{original\ number} \times 100 =$

$\frac{25-50}{50} \times 100 = \frac{-25}{50} \times 100 = -50$. The percentage decrease is 50.

(the negative sign means percentage decrease)

Therefore, the price of the table decreased by 50%.

Discount, Tax and Tip

☆ To find the discount: Multiply the regular price by the rate of discount

☆ To find the selling price: Original price − discount

☆ To find tax: Multiply the tax rate to the taxable amount (income, property value, etc.)

☆ To find the tip, multiply the rate to the selling price.

Examples:

Example 1. With an 25% discount, Ella saved $50 on a dress. What was the original price of the dress?

Solution: let x be the original price of the dress. Then: 25 % of $x = 50$. Write an equation and solve for x: $0.25 \times x = 50 \rightarrow x = \frac{50}{0.25} = 200$. The original price of the dress was $200.

Example 2. Sophia purchased a new computer for a price of $820 at the Apple Store. What is the total amount her credit card is charged if the sales tax is 10%?

Solution: The taxable amount is $820, and the tax rate is 10%. Then:
$$Tax = 0.10 \times 820 = 82$$
$Final\ price = Selling\ price + Tax \rightarrow final\ price = \$820 + \$82 = \$902$

Example 3. Nicole and her friends went out to eat at a restaurant. If their bill was $50 and they gave their server a 12% tip, how much did they pay altogether?

Solution: First, find the tip. To find the tip, multiply the rate to the bill amount. $Tip = 50 \times 0.12 = 6$. The final amount is: $50 + $6 = $56

Simple Interest

★ Simple Interest: The charge for borrowing money or the return for lending it.

★ Simple interest is calculated on the initial amount (principal).

★ To solve a simple interest problem, use this formula:

$$Interest = principal \times rate \times time \rightarrow (I = p \times r \times t = prt)$$

Examples:

Example 1. Find simple interest for $250 investment at 6% for 5 years.

Solution: Use Interest formula:
$I = prt$ ($P = \$250, r = 6\% = \frac{6}{100} = 0.06$ and $t = 5$)
Then: $I = 250 \times 0.06 \times 5 = \$75$

Example 2. Find simple interest for $1,300 at 3% for 2 years.

Solution: Use Interest formula:
$I = prt$ ($P = \$1,300, r = 3\% = \frac{3}{100} = 0.03$ and $t = 2$)
Then: $I = 1,300 \times 0.03 \times 2 = \$78.00$

Example 3. Andy received a student loan to pay for his educational expenses this year. What is the interest on the loan if he borrowed $5,100 at 5% for 4 years?

Solution: Use Interest formula: $I = prt$. $P = \$5,100$, $r = 5\% = 0.05$ and $t = 4$
Then: $I = 5,100 \times 0.05 \times 4 = \$1,020$

Example 4. Bob is starting his own small business. He borrowed $28,000 from the bank at an 7% rate for 6 months. Find the interest Bob will pay on this loan.

Solution: Use Interest formula:
$I = prt$. $P = \$28,000$, $r = 7\% = 0.07$ and $t = 0.5$ (6 months is half year). Then:
$I = 28,000 \times 0.07 \times 0.5 = \$980$

Day 3: Practices

✏ Reduce each ratio.

1) $3:15 = \_\_\_:\_\_\_$

2) $8:72 = \_\_\_:\_\_\_$

3) $15:25 = \_\_\_:\_\_\_$

4) $35:10 = \_\_\_:\_\_\_$

5) $36:42 = \_\_\_:\_\_\_$

6) $24:64 = \_\_\_:\_\_\_$

✏ Solve.

7) In Jack's class, 48 of the students are tall and 20 are short. In Michael's class 28 students are tall and 12 students are short. Which class has a higher ratio of tall to short students? _____

8) You can buy 7 cans of green beans at a supermarket for $7.49. How much does it cost to buy 21 cans of green beans? _____

✏ Solve each proportion.

9) $\frac{3}{4} = \frac{12}{x} \rightarrow x = \_\_\_\_$

10) $\frac{2}{5} = \frac{x}{20} \rightarrow x = \_\_\_\_$

11) $\frac{9}{x} = \frac{3}{5} \rightarrow x = \_\_\_\_$

12) $\frac{x}{24} = \frac{5}{6} \rightarrow x = \_\_\_\_$

13) $\frac{8}{4} = \frac{x}{16} \rightarrow x = \_\_\_\_$

14) $\frac{3}{x} = \frac{12}{16} \rightarrow x = \_\_\_\_$

15) $\frac{24}{32} = \frac{3}{x} \rightarrow x = \_\_\_\_$

16) $\frac{x}{7} = \frac{21}{49} \rightarrow x = \_\_\_\_$

✏ Solve each problem.

17) Two rectangles are similar. The first is 6 *feet* wide and 36 *feet* long. The second is 10 *feet* wide. What is the length of the second rectangle? _____

18) Two rectangles are similar. One is 4.6 *meters* by 7 *meters*. The longer side of the second rectangle is 28 *meters*. What is the other side of the second rectangle? _____

Effortless Math Education

Solve each problem.

19) What is 15% of 60? ____

20) What is 20% of 500? ____

21) 25 what is percent of 250? ____

22) 30 is what percent of 150? ____

23) 15 is 10 percent of what number? ____

24) 25 is 5 percent of what number? ____

Solve each problem.

25) Bob got a raise, and his hourly wage increased from $15 to $21. What is the percent increase? ____ %

26) A $45 shirt now selling for $36 is discounted by what percent? ____ %

Find the selling price of each item.

27) Original price of a computer: $500

    Tax: 5%, Selling price: $_____

28) Nicolas hired a moving company. The company charged $500 for its services, and Nicolas gives the movers a 14% tip. How much does Nicolas tip the movers? $____

29) Mason has lunch at a restaurant and the cost of his meal is $60. Mason wants to leave a 10% tip. What is Mason's total bill, including tip? $_____

Determine the simple interest for the following loans.

30) $800 at 3% for 2 years. $____

31) $260 at 10% for 5 years. $____

32) $380 at 4% for 3 years. $____

33) $1,200 at 2% for 1 years. $____

Effortless Math Education

EffortlessMath.com

Day 3: Answers

1) $3:15 \rightarrow 3 \div 3 = 1, \ 15 \div 3 = 5 \rightarrow 3:15 = 1:5$

2) $8:72 \rightarrow 8 \div 8 = 1, \ 72 \div 8 = 9 \rightarrow 8:72 = 1:9$

3) $15:25 \rightarrow 15 \div 5 = 3, \ 25 \div 5 = 5 \rightarrow 15:25 = 3:5$

4) $35:10 \rightarrow 35 \div 5 = 7, \ 10 \div 5 = 2 \rightarrow 35:10 = 7:2$

5) $36:42 \rightarrow 36 \div 6 = 6, \ 42 \div 6 = 7 \rightarrow 36:42 = 6:7$

6) $24:64 \rightarrow 24 \div 8 = 3, \ 64 \div 8 = 8 \rightarrow 24:64 = 3:8$

7) In Jack's class the ratio of tall students to short students is: $\frac{48}{20} = \frac{48 \div 4}{20 \div 4} = \frac{12}{5}$ and in Michael's class the ratio is: $\frac{28}{12} = \frac{28 \div 4}{12 \div 4} = \frac{7}{3}$. Compare two fractions: $\frac{12}{5} = \frac{12 \times 3}{5 \times 3} = \frac{36}{15}$ and $\frac{7}{3} = \frac{7 \times 5}{3 \times 5} = \frac{35}{15} \rightarrow \frac{36}{15} > \frac{35}{15}$. In Jack's class the ratio of tall to short students is higher.

8) Write a proportion and solve: $\frac{7}{7.49} = \frac{21}{x} \rightarrow x = 3 \times 7.49 = \$22.47$

9) $\frac{3}{4} = \frac{12}{x} \rightarrow 3 \times x = 4 \times 12 \rightarrow 3x = 48 \rightarrow x = \frac{48}{3} = 16$

10) $\frac{2}{5} = \frac{x}{20} \rightarrow 5 \times x = 2 \times 20 \rightarrow 5x = 40 \rightarrow x = \frac{40}{5} = 8$

11) $\frac{9}{x} = \frac{3}{5} \rightarrow 9 \times 5 = 3 \times x \rightarrow 45 = 3x \rightarrow x = \frac{45}{3} = 15$

12) $\frac{x}{24} = \frac{5}{6} \rightarrow 6 \times x = 5 \times 24 \rightarrow 6x = 120 \rightarrow x = \frac{120}{6} = 20$

13) $\frac{8}{4} = \frac{x}{16} \rightarrow 8 \times 16 = 4 \times x \rightarrow 128 = 4x \rightarrow x = \frac{128}{4} = 32$

14) $\frac{3}{x} = \frac{12}{16} \rightarrow 3 \times 16 = 12 \times x \rightarrow 48 = 12x \rightarrow x = \frac{48}{12} = 4$

15) $\frac{24}{32} = \frac{3}{x} \rightarrow 24 \times x = 3 \times 32 \rightarrow 24x = 96 \rightarrow x = \frac{96}{24} = 4$

16) $\frac{x}{7} = \frac{21}{49} \rightarrow 49 \times x = 7 \times 21 \rightarrow 49x = 147 \rightarrow x = \frac{147}{49} = 3$

17) $\frac{6}{10} = \frac{36}{x} \rightarrow 6 \times x = 36 \times 10 \rightarrow 6x = 360 \rightarrow x = \frac{360}{6} = 60 \rightarrow x = 60 \ feet$

18) $\frac{4.6}{7} = \frac{x}{28} \rightarrow 7 \times x = 28 \times 4.6 \rightarrow 7x = 128.8 \rightarrow x = \frac{128.8 \div 7}{7 \div 7} = 18.4 \rightarrow x = 18.4 \ meters$

19) Part = Percent × Base → $15\% \times 60 = \frac{15}{100} \times 60 = 0.15 \times 60 = 9$

20) Part = Percent × Base → $20\% \times 500 = \frac{20}{100} \times 500 = 0.2 \times 500 = 100$

21) Percent = Part ÷ Base → $25 \div 250 = \frac{25}{250} = \frac{25 \div 25}{250 \div 25} = \frac{1}{10} = \frac{1}{10} \times 100 = 10\%$

22) Percent = Part ÷ Base → $30 \div 150 = \frac{30}{150} = \frac{30 \div 30}{150 \div 30} = \frac{1}{5} = \frac{1}{5} \times 100 = 20\%$

23) Base = Part ÷ Percent → $15 \div 10\% = 15 \div \frac{10}{100} = 15 \div 0.1 = 150$

24) Base = Part ÷ Percent → $25 \div 5\% = 25 \div \frac{5}{100} = 25 \div 0.05 = 500$

25) Percent of change = $\frac{new\ number - original\ number}{original\ number} \times 100 = \frac{21-15}{15} \times 100 = \frac{6}{15} \times 100 = 40\%$

26) $\frac{36-45}{45} \times 100 = \frac{-9}{45} \times 100 = -20\%$ (the negative sign means that the price decreased)

27) $5\% \times 500 = \frac{5}{100} \times 500 = 25 \rightarrow \$500 + \$25 = \$525$

28) $14\% \times \$500 = \frac{14}{100} \times \$500 = \$500 \times 0.14 = \$70$

29) $10\% \times \$60 = \frac{10}{100} \times \$60 = \$6 \rightarrow \$60 + \$6 = \$66$

30) $800 \times 3\% \times 2 = 800 \times \frac{3}{100} \times 2 = \frac{4,800}{100} = 48$

31) $260 \times 10\% \times 5 = 260 \times \frac{10}{100} \times 5 = 260 \times \frac{1}{10} \times 5 = 130$

32) $380 \times 4\% \times 3 = 380 \times \frac{4}{100} \times 3 = 45.6$

33) $1,200 \times 2\% \times 1 = 1,200 \times \frac{2}{100} \times 1 = 24$

Day 4: Exponents and Variables

Math topics that you'll learn in this chapter:

1. Multiplication Property of Exponents
2. Division Property of Exponents
3. Powers of Products and Quotients
4. Zero and Negative Exponents
5. Negative Exponents and Negative Bases
6. Scientific Notation
7. Radicals

Multiplication Property of Exponents

☆ Exponents are shorthand for repeated multiplication of the same number by itself. For example, instead of 2×2, we can write 2^2. For $3 \times 3 \times 3 \times 3$, we can write 3^4.

☆ In algebra, a variable is a letter used to stand for a number. The most common letters are: $x, y, z, a, b, c, m,$ and n.

☆ Exponent's rules: $x^a \times x^b = x^{a+b}$, $\frac{x^a}{x^b} = x^{a-b}$

$(x^a)^b = x^{a \times b}$ $\qquad$ $(xy)^a = x^a \times y^a$ $\qquad$ $\left(\frac{a}{b}\right)^c = \frac{a^c}{b^c}$

Examples:

Example 1. Multiply. $3x^2 \times 4x^3$

Solution: Use Exponent's rules: $x^a \times x^b = x^{a+b} \rightarrow x^2 \times x^3 = x^{2+3} = x^5$
Then: $3x^2 \times 4x^3 = 12x^5$

Example 2. Simplify. $(x^2 y^4)^3$

Solution: Use Exponent's rules: $(x^a)^b = x^{a \times b}$.
Then: $(x^2 y^4)^3 = x^{2 \times 3} y^{4 \times 3} = x^6 y^{12}$

Example 3. Multiply. $6x^7 \times 4x^3$

Solution: Use Exponent's rules: $x^a \times x^b = x^{a+b} \rightarrow x^7 \times x^3 = x^{7+3} = x^{10}$
Then: $6x^7 \times 4x^3 = 24x^{10}$

Example 4. Simplify. $(x^2 y^5)^4$

Solution: Use Exponent's rules: $(x^a)^b = x^{a \times b}$.
Then: $(x^2 y^5)^4 = x^{2 \times 4} y^{5 \times 4} = x^8 y^{20}$

Division Property of Exponents

For division of exponents use following formulas:

☆ $\frac{x^a}{x^b} = x^{a-b}$ $(x \neq 0)$

☆ $\frac{x^a}{x^b} = \frac{1}{x^{b-a}}$, $(x \neq 0)$

☆ $\frac{1}{x^b} = x^{-b}$

Examples:

Example 1. Simplify. $\frac{12x^2y}{3xy^3} =$

Solution: First, cancel the common factor: $3 \rightarrow \frac{12x^2y}{3xy^3} = \frac{4x^2y}{xy^3}$

Use Exponent's rules: $\frac{x^a}{x^b} = x^{a-b} \rightarrow \frac{x^2}{x} = x^{2-1} = x^1$ and $\frac{x^a}{x^b} = \frac{1}{x^{b-a}} \rightarrow \frac{y}{y^3} = \frac{1}{y^{3-1}} = \frac{1}{y^2}$

Then: $\frac{12x^2y}{3xy^3} = \frac{4x}{y^2}$

Example 2. Simplify. $\frac{48x^{12}}{16x^9} =$

Solution: Use Exponent's rules: $\frac{x^a}{x^b} = x^{a-b} \rightarrow \frac{x^{12}}{x^9} = x^{12-9} = x^3$

Then: $\frac{48x^{12}}{16x^9} = 3x^3$

Example 3. Simplify. $\frac{6x^5y^7}{42x^8y^2} =$

Solution: First, cancel the common factor: $6 \rightarrow \frac{x^5y^7}{7x^8y^2}$

Use Exponent's rules: $\frac{x^a}{x^b} = x^{a-b} \rightarrow \frac{x^5}{x^8} = x^{5-8} = x^{-3} = \frac{1}{x^3}$ and $\frac{y^7}{y^2} = y^{7-2} = y^5$

Then: $\frac{6x^5y^7}{42x^8y^2} = \frac{y^5}{7x^3}$

Powers of Products and Quotients

☆ For any nonzero numbers a and b and any integer x, $(ab)^x = a^x \times b^x$ and $\left(\frac{a}{b}\right)^c = \frac{a^c}{b^c}$

Examples:

Example 1. Simplify. $(6x^4y^5)^2$

Solution: Use Exponent's rules: $(x^a)^b = x^{a \times b}$

$(6x^4y^5)^2 = (6)^2(x^4)^2(y^5)^2 = 36x^{4 \times 2}y^{5 \times 2} = 36x^8y^{10}$

Example 2. Simplify. $\left(\frac{5x^5}{2x^4}\right)^2$

Solution: First, cancel the common factor: $x^4 \to \left(\frac{5x^5}{2x^4}\right) = \left(\frac{5x}{2}\right)^2$

Use Exponent's rules: $\left(\frac{a}{b}\right)^c = \frac{a^c}{b^c}$. Then: $\left(\frac{5x}{2}\right)^2 = \frac{(5x)^2}{(2)^2} = \frac{25x^2}{4}$

Example 3. Simplify. $(-6x^7y^3)^2$

Solution: Use Exponent's rules: $(x^a)^b = x^{a \times b}$

$(-6x^7y^3)^2 = (-6)^2(x^7)^2(y^3)^2 = 36x^{7 \times 2}y^{3 \times 2} = 36x^{14}y^6$

Example 4. Simplify. $\left(\frac{8x^3}{5x^7}\right)^2$

Solution: First, cancel the common factor: $x^3 \to \left(\frac{8x^3}{5x^7}\right)^2 = \left(\frac{8}{5x^4}\right)^2$

Use Exponent's rules: $\left(\frac{a}{b}\right)^c = \frac{a^c}{b^c}$, Then: $\left(\frac{8}{5x^4}\right)^2 = \frac{8^2}{(5x^4)^2} = \frac{64}{25x^8}$

Zero and Negative Exponents

★ Zero-Exponent Rule: $a^0 = 1$, this means that anything raised to the zero power is 1. For example: $(5xy)^0 = 1$ (number zero is an exception: $0^0 = 0$)

★ A negative exponent simply means that the base is on the wrong side of the fraction line, so you need to flip the base to the other side. For instance, "x^{-2}" (pronounced as "ecks to the minus two") just means "x^2" but underneath, as in $\frac{1}{x^2}$.

Examples:

Example 1. Evaluate. $\left(\frac{3}{7}\right)^{-2} =$

Solution: Use negative exponent's rule: $\left(\frac{x^a}{x^b}\right)^{-2} = \left(\frac{x^b}{x^a}\right)^2 \to \left(\frac{3}{7}\right)^{-2} = \left(\frac{7}{3}\right)^2$
Then: $\left(\frac{7}{3}\right)^2 = \frac{7^2}{3^2} = \frac{49}{9}$

Example 2. Evaluate. $\left(\frac{2}{5}\right)^{-3} =$

Solution: Use negative exponent's rule: $\left(\frac{x^a}{x^b}\right)^{-3} = \left(\frac{x^b}{x^a}\right)^3 \to \left(\frac{2}{5}\right)^{-3} = \left(\frac{5}{2}\right)^3 =$
Then: $\left(\frac{5}{2}\right)^3 = \frac{5^3}{2^3} = \frac{125}{8}$

Example 3. Evaluate. $\left(\frac{a}{b}\right)^0 =$

Solution: Use zero-exponent Rule: $a^0 = 1$
Then: $\left(\frac{a}{b}\right)^0 = 1$

Example 4. Evaluate. $\left(\frac{9}{5}\right)^{-1} =$

Solution: Use negative exponent's rule: $\left(\frac{x^a}{x^b}\right)^{-1} = \left(\frac{x^b}{x^a}\right)^1 \to \left(\frac{9}{5}\right)^{-1} = \left(\frac{5}{9}\right)^1 = \frac{5}{9}$

Negative Exponents and Negative Bases

☆ A negative exponent is the reciprocal of that number with a positive exponent. $(3)^{-2} = \frac{1}{3^2}$

☆ To simplify a negative exponent, make the power positive!

☆ The parenthesis is important! -5^{-2} is not the same as $(-5)^{-2}$

$-5^{-2} = -\frac{1}{5^2}$ and $(-5)^{-2} = +\frac{1}{5^2}$

Examples:

Example 1. Simplify. $\left(\frac{4a}{7c}\right)^{-2} =$

Solution: Use negative exponent's rule: $\left(\frac{x^a}{x^b}\right)^{-2} = \left(\frac{x^b}{x^a}\right)^2 \rightarrow \left(\frac{4a}{7c}\right)^{-2} = \left(\frac{7c}{4a}\right)^2$

Now use exponent's rule: $\left(\frac{a}{b}\right)^c = \frac{a^c}{b^c} \rightarrow \left(\frac{7c}{4a}\right)^2 = \frac{7^2 c^2}{4^2 a^2}$

Then: $\frac{7^2 c^2}{4^2 a^2} = \frac{49 c^2}{16 a^2}$

Example 2. Simplify. $\left(\frac{3x}{y}\right)^{-3} =$

Solution: Use negative exponent's rule: $\left(\frac{x^a}{x^b}\right)^{-3} = \left(\frac{x^b}{x^a}\right)^3 \rightarrow \left(\frac{3x}{y}\right)^{-3} = \left(\frac{y}{3x}\right)^3$

Now use exponent's rule: $\left(\frac{a}{b}\right)^c = \frac{a^c}{b^c} \rightarrow \left(\frac{y}{3x}\right)^3 = \frac{y^3}{3^3 x^3} = \frac{y^3}{27 x^3}$

Example 3. Simplify. $\left(\frac{7a}{4c}\right)^{-2} =$

Solution: Use negative exponent's rule: $\left(\frac{x^a}{x^b}\right)^{-2} = \left(\frac{x^b}{x^a}\right)^2 \rightarrow \left(\frac{7a}{4c}\right)^{-2} = \left(\frac{4c}{7a}\right)^2$

Now use exponent's rule: $\left(\frac{a}{b}\right)^c = \frac{a^c}{b^c} \rightarrow \left(\frac{4c}{7a}\right)^2 = \frac{4^2 c^2}{7^2 a^2}$

Then: $\frac{4^2 c^2}{7^2 a^2} = \frac{16 c^2}{49 a^2}$

Scientific Notation

★ Scientific notation is used to write very big or very small numbers in decimal form.

★ In scientific notation, all numbers are written in the form of: $m \times 10^n$, where m is greater than 1 and less than 10.

★ To convert a number from scientific notation to standard form, move the decimal point to the left (if the exponent of ten is a negative number), or to the right (if the exponent is positive).

Examples:

Example 1. *Write* 0.000037 *in scientific notation.*

Solution: First, move the decimal point to the right so you have a number between 1 and 10. That number is 3.7. Now, determine how many places the decimal moved in step 1 by the power of 10. We moved the decimal point 5 digits to the right. Then: 10^{-5}. When the decimal moved to the right, the exponent is negative. Then: $0.000037 = 3.7 \times 10^{-5}$

Example 2. *Write* 5.3×10^{-3} *in standard notation.*

Solution: The exponent is negative 3. Then, move the decimal point to the left three digits. (remember $5.3 = 0000005.3$) When the decimal moved to the right, the exponent is negative. Then: $5.3 \times 10^{-3} = 0.0053$

Example 3. *Write* 0.00042 *in scientific notation.*

Solution: First, move the decimal point to the right so you have a number between 1 and 10. Then: $m = 4.2$. Now, determine how many places the decimal moved in step 1 by the power of 10. 10^{-4}. Then: $0.00042 = 4.2 \times 10^{-4}$

Example 4. *Write* 7.3×10^7 *in standard notation.*

Solution: The exponent is positive 7. Then, move the decimal point to the right seven digits. (remember $7.3 = 7.3000000$) Then: $7.3 \times 10^7 = 73,000,000$

# Radicals

☆ If n is a positive integer and x is a real number, then: $\sqrt[n]{x} = x^{\frac{1}{n}}$, $\sqrt[n]{xy} = x^{\frac{1}{n}} \times y^{\frac{1}{n}}$, $\sqrt[n]{\frac{x}{y}} = \frac{x^{\frac{1}{n}}}{y^{\frac{1}{n}}}$, and $\sqrt[n]{x} \times \sqrt[n]{y} = \sqrt[n]{xy}$

☆ A square root of x is a number r whose square is: $r^2 = x$ (r is a square root of x)

☆ To add and subtract radicals, we need to have the same values under the radical. For example: $\sqrt{5} + 3\sqrt{5} = 4\sqrt{5}$, $5\sqrt{6} - 2\sqrt{6} = 3\sqrt{5}$

Examples:

Example 1. Find the square root of $\sqrt{256}$.

Solution: First, factor the number: $256 = 16^2$. Then: $\sqrt{256} = \sqrt{16^2}$
Now use radical rule: $\sqrt[n]{a^n} = a$. Then: $\sqrt{256} = \sqrt{16^2} = 16$

Example 2. Evaluate. $\sqrt{9} \times \sqrt{36} =$

Solution: Find the values of $\sqrt{9}$ and $\sqrt{36}$. Then: $\sqrt{9} \times \sqrt{36} = 3 \times 6 = 18$

Example 3. Solve. $3\sqrt{7} + 11\sqrt{7}$.

Solution: Since we have the same values under the radical, we can add these two radicals: $3\sqrt{7} + 11\sqrt{7} = 14\sqrt{7}$

Example 4. Evaluate. $\sqrt{40} \times \sqrt{10} =$

Solution: Use this radical rule: $\sqrt[n]{x} \times \sqrt[n]{y} = \sqrt[n]{xy} \to \sqrt{40} \times \sqrt{10} = \sqrt{400}$
The square root of 400 is 20. Then: $\sqrt{40} \times \sqrt{10} = \sqrt{400} = 20$

Day 4: Practices

✎ Find the products.

1) $5x^3 \times 2x =$

2) $x^4 \times 5x^2y =$

3) $2xy \times 3x^5y^2 =$

4) $4xy^2 \times 2x^2y =$

5) $-3x^3y^3 \times 2x^2y^2 =$

6) $-5xy^2 \times 3x^5y^2 =$

7) $-5x^2y^6 \times 6x^5y^2 =$

8) $-2x^3y^3 \times 2x^3y^3 =$

9) $-7xy^3 \times 4x^5y^2 =$

10) $-x^4y^3 \times (-5x^6y^2) =$

11) $-6y^6 \times 7x^6y^2 =$

12) $-8x^4 \times 2y^2 =$

✎ Simplify.

13) $\dfrac{3^2 \times 3^3}{3^3 \times 3} =$

14) $\dfrac{4^2 \times 4^4}{5^4 \times 5} =$

15) $\dfrac{14x^5}{7x^2} =$

16) $\dfrac{15x^3}{5x^6} =$

17) $\dfrac{64y^3}{8xy^7} =$

18) $\dfrac{10x^4y^5}{30x^5y^4} =$

19) $\dfrac{11y}{44x^3y^3} =$

20) $\dfrac{40xy^3}{120xy^3} =$

21) $\dfrac{45x^3}{25xy^3} =$

22) $\dfrac{72y^6x}{36x^8y^9} =$

✎ Solve.

23) $(x^2 y^2)^3 =$

24) $(2x^3 y^2)^3 =$

25) $(2x \times 3xy^2)^2 =$

26) $(4x \times 2y^4)^2 =$

27) $\left(\dfrac{3x}{x^2}\right)^2 =$

28) $\left(\dfrac{6y}{18y^3}\right)^2 =$

29) $\left(\dfrac{3x^2y^2}{12x^4y^3}\right)^3 =$

30) $\left(\dfrac{23x^5y^3}{46x^3y^5}\right)^3 =$

31) $\left(\dfrac{16x^7y^3}{48x^5y^2}\right)^2 =$

32) $\left(\dfrac{12x^5y^6}{60x^7y^2}\right)^2 =$

Evaluate each expression. (Zero and Negative Exponents)

33) $\left(\frac{1}{3}\right)^{-2} =$

34) $\left(\frac{1}{4}\right)^{-3} =$

35) $\left(\frac{1}{6}\right)^{-2} =$

36) $\left(\frac{2}{3}\right)^{-3} =$

37) $\left(\frac{2}{5}\right)^{-3} =$

38) $\left(\frac{3}{5}\right)^{-2} =$

Write each expression with positive exponents.

39) $2y^{-3} =$

40) $13y^{-5} =$

41) $-20x^{-2} =$

42) $15a^{-2}b^3 =$

43) $23a^2b^{-4}c^{-8} =$

44) $-4x^4y^{-2} =$

45) $\frac{16y}{x^3 y^{-4}} =$

46) $\frac{30a^{-3}b}{-100c^{-2}} =$

Write each number in scientific notation.

47) $0.00518 =$

48) $0.000042 =$

49) $78,000 =$

50) $92,000,000 =$

Evaluate.

51) $\sqrt{5} \times \sqrt{5} =$

52) $\sqrt{25} - \sqrt{4} =$

53) $\sqrt{81} + \sqrt{36} =$

54) $\sqrt{4} \times \sqrt{25} =$

55) $\sqrt{2} \times \sqrt{18} =$

56) $4\sqrt{2} + 3\sqrt{2} =$

57) $5\sqrt{7} + 2\sqrt{7} =$

58) $\sqrt{45} + 2\sqrt{5} =$

Day 4: Answers

1) $5x^3 \times 2x \to x^3 \times x^1 = x^{3+1} = x^4 \to 5x^3 \times 2x = 10x^4$
2) $x^4 \times 5x^2y \to x^4 \times x^2 = x^{4+2} = x^6 \to x^4 \times 5x^2y = 5x^6y$
3) $2xy \times 3x^5y^2 \to x \times x^5 = x^{1+5} = x^6,\ y \times y^2 = y^{1+2} = y^3 \to 2xy \times 3x^5y^2 = 6x^6y^3$
4) $4xy^2 \times 2x^2y \to x \times x^2 = x^{1+2} = x^3,\ y^2 \times y = y^{2+1} = y^3 \to 4xy^2 \times 2x^2y = 8x^3y^3$
5) $-3x^3y^3 \times 2x^2y^2 \to x^3 \times x^2 = x^{3+2} = x^5,\ y^3 \times y^2 = y^{3+2} = y^5 \to$
    $-3x^3y^3 \times 2x^2y^2 = -6x^5y^5$
6) $-5xy^2 \times 3x^5y^2 \to x \times x^5 = x^{1+5} = x^6,\ y^2 \times y^2 = y^{2+2} = y^4 \to$
    $-5xy^2 \times 3x^5y^2 = -15x^6y^4$
7) $-5x^2y^6 \times 6x^5y^2 \to x^2 \times x^5 = x^{2+5} = x^7,\ y^6 \times y^2 = y^{6+2} = y^8 \to$
    $-5x^2y^6 \times 6x^5y^2 = -30x^7y^8$
8) $-2x^3y^3 \times 2x^3y^3 \to x^3 \times x^3 = x^{3+3} = x^6,\ y^3 \times y^3 = y^{3+3} = y^6 \to$
    $-2x^3y^3 \times 2x^3y^3 = -4x^6y^6$
9) $-7xy^3 \times 4x^5y^2 \to x \times x^5 = x^{1+5} = x^6,\ y^3 \times y^2 = y^{3+2} = y^5 \to$
    $-7xy^3 \times 4x^5y^2 = -28x^6y^5$
10) $-x^4y^3 \times (-5x^6y^2) \to x^4 \times x^6 = x^{4+6} = x^{10},\ y^3 \times y^2 = y^{3+2} = y^5 \to$
    $-x^4y^3 \times (-5x^6y^2) = 5x^{10}y^5$
11) $-6y^6 \times 7x^6y^2 \to y^6 \times y^2 = y^{6+2} = y^8 \to -6y^6 \times 7x^6y^2 = -42x^6y^8$
12) $-8x^4 \times 2y^2 = -16x^4y^2$
13) $\frac{3^2 \times 3^3}{3^3 \times 3} = \frac{3^{2+3}}{3^{3+1}} = \frac{3^5}{3^4} = 3^{5-4} = 3^1 = 3$
14) $\frac{4^2 \times 4^4}{5^4 \times 5} = \frac{4^{2+4}}{5^{4+1}} = \frac{4^6}{5^5}$
15) $\frac{14x^5}{7x^2} \to \frac{14 \div 7}{7 \div 7} = 2,\ \frac{x^5}{x^2} = x^{5-2} = x^3 \to \frac{14x^5}{7x^2} = 2x^3$
16) $\frac{15x^3}{5x^6} \to \frac{15 \div 5}{5 \div 5} = 3,\ \frac{x^3}{x^6} = x^{3-6} = x^{-3} = \frac{1}{x^3} \to \frac{15x^3}{5x^6} = \frac{3}{x^3}$
17) $\frac{64y^3}{8xy^7} \to \frac{64 \div 8}{8 \div 8} = 8,\ \frac{y^3}{y^7} = y^{3-7} = y^{-4} = \frac{1}{y^4} \to \frac{64y^3}{8xy^7} = \frac{8}{xy^4}$
18) $\frac{10x^4y^5}{30x^5y^4} \to \frac{10 \div 10}{30 \div 10} = \frac{1}{3},\ \frac{x^4}{x^5} = x^{4-5} = x^{-1} = \frac{1}{x},\ \frac{y^5}{y^4} = y^{5-4} = y^1 \to \frac{10x^4y^5}{30x^5y^4} = \frac{y}{3x}$
19) $\frac{11y}{44x^3y^3} \to \frac{11 \div 11}{44 \div 11} = \frac{1}{4},\ \frac{y}{y^3} = y^{1-3} = y^{-2} = \frac{1}{y^2} \to \frac{11y}{44x^3y^3} = \frac{1}{4x^3y^2}$
20) $\frac{40xy^3}{120xy^3} \to \frac{40 \div 40}{120 \div 40} = \frac{1}{3},\ \frac{x}{x} = x^{1-1} = x^0 = 1,\ \frac{y^3}{y^3} = y^{3-3} = y^0 = 1 \to \frac{40xy^3}{120xy^3} = \frac{1}{3}$

Day 4: Answers — Exponents and Variables

21) $\dfrac{45x^3}{25xy^3} \rightarrow \dfrac{45 \div 5}{25 \div 5} = \dfrac{9}{5}, \dfrac{x^3}{x} = x^{3-1} = x^2 \rightarrow \dfrac{45x^3}{25xy^3} = \dfrac{9x^2}{5y^3}$

22) $\dfrac{72y^6 x}{36x^8 y^9} \rightarrow \dfrac{72 \div 36}{36 \div 36} = \dfrac{2}{1} = 2, \dfrac{y^6}{y^9} = y^{6-9} = y^{-3} = \dfrac{1}{y^3}, \dfrac{x}{x^8} = x^{1-8} = x^{-7} = \dfrac{1}{x^7} \rightarrow \dfrac{72y^6 x}{36x^8 y^9} = \dfrac{2}{x^7 y^3}$

23) $(x^2 y^2)^3 = x^{2\times 3} y^{2\times 3} = x^6 y^6$

24) $(2x^3 y^2)^3 = 2^3 x^{3\times 3} y^{2\times 3} = 8x^9 y^6$

25) $(2x \times 3xy^2)^2 = (2 \times 3)^2 x^{(1+1)\times 2} y^{2\times 2} = 6^2 x^{2\times 2} y^4 = 36x^4 y^4$

26) $(4x \times 2y^4)^2 = (4 \times 2)^2 x^{1\times 2} y^{4\times 2} = 8^2 x^2 y^8 = 64x^2 y^8$

27) $\left(\dfrac{3x}{x^2}\right)^2 = \dfrac{3^{1\times 2} x^{1\times 2}}{x^{2\times 2}} = \dfrac{3^2 x^2}{x^4} \rightarrow 3^2 = 9, \dfrac{x^2}{x^4} = x^{2-4} = x^{-2} = \dfrac{1}{x^2} \rightarrow \left(\dfrac{3x}{x^2}\right)^2 = \dfrac{9}{x^2}$

28) $\left(\dfrac{6y}{18y^3}\right)^2 = \dfrac{6^{1\times 2} y^{1\times 2}}{(18)^{1\times 2} y^{3\times 2}} = \dfrac{6^2 y^2}{18^2 y^6} \rightarrow \dfrac{36}{324} = \dfrac{1}{9}, \dfrac{y^2}{y^6} = y^{2-6} = y^{-4} = \dfrac{1}{y^4} \rightarrow \left(\dfrac{6y}{18y^3}\right)^2 = \dfrac{1}{9y^4}$

29) $\left(\dfrac{3x^2 y^2}{12x^4 y^3}\right)^3 = \dfrac{3^{1\times 3} x^{2\times 3} y^{2\times 3}}{(12)^{1\times 3} x^{4\times 3} y^{3\times 3}} = \dfrac{3^3 x^6 y^6}{12^3 x^{12} y^9} \rightarrow \dfrac{3^3}{12^3} = \dfrac{27}{1{,}728} = \dfrac{1}{64}, \dfrac{x^6}{x^{12}} = x^{6-12} = x^{-6} = \dfrac{1}{x^6}, \dfrac{y^6}{y^9} = y^{6-9} = y^{-3} = \dfrac{1}{y^3} \rightarrow \left(\dfrac{3x^2 y^2}{12x^4 y^3}\right)^3 = \dfrac{1}{64x^6 y^3}$

30) $\left(\dfrac{23x^5 y^3}{46x^3 y^5}\right)^3 = \dfrac{23^{1\times 3} x^{5\times 3} y^{3\times 3}}{(23\times 2)^{1\times 3} x^{3\times 3} y^{5\times 3}} = \dfrac{23^3 x^{15} y^9}{23^3 \times 2^3 x^9 y^{15}} = \dfrac{x^{15} y^9}{8x^9 y^{15}} \rightarrow \dfrac{x^{15}}{x^9} = x^{15-9} = x^6, \dfrac{y^9}{y^{15}} = y^{9-15} = y^{-6} = \dfrac{1}{y^6} \rightarrow \left(\dfrac{23x^5 y^3}{46x^3 y^5}\right)^3 = \dfrac{x^6}{8y^6}$

31) $\left(\dfrac{16x^7 y^3}{48x^5 y^2}\right)^2 = \dfrac{16^{1\times 2} x^{7\times 2} y^{3\times 2}}{(16\times 3)^{1\times 2} x^{5\times 2} y^{2\times 2}} = \dfrac{16^2 x^{14} y^6}{16^2 \times 3^2 x^{10} y^4} = \dfrac{x^{14} y^6}{9x^{10} y^4} \rightarrow \dfrac{x^{14}}{x^{10}} = x^{14-10} = x^4, \dfrac{y^6}{y^4} = y^{6-4} = y^2 \rightarrow \left(\dfrac{16x^7 y^3}{48x^5 y^2}\right)^2 = \dfrac{x^4 y^2}{9}$

32) $\left(\dfrac{12x^5 y^6}{60x^7 y^2}\right)^2 = \dfrac{12^{1\times 2} x^{5\times 2} y^{6\times 2}}{(12\times 5)^{1\times 2} x^{7\times 2} y^{2\times 2}} = \dfrac{12^2 x^{10} y^{12}}{12^2 \times 5^2 x^{14} y^4} = \dfrac{x^{10} y^{12}}{25 x^{14} y^4} \rightarrow \dfrac{x^{10}}{x^{14}} = x^{10-14} = x^{-4} = \dfrac{1}{x^4}, \dfrac{y^{12}}{y^4} = y^{12-4} = y^8 \rightarrow \left(\dfrac{12x^5 y^6}{60x^7 y^2}\right)^2 = \dfrac{y^8}{25x^4}$

33) $\left(\dfrac{1}{3}\right)^{-2} = 3^2 = 9$

34) $\left(\dfrac{1}{4}\right)^{-3} = 4^3 = 64$

35) $\left(\dfrac{1}{6}\right)^{-2} = 6^2 = 36$

36) $\left(\dfrac{2}{3}\right)^{-3} = \left(\dfrac{3}{2}\right)^3 = \dfrac{3^3}{2^3} = \dfrac{27}{8}$

37) $\left(\dfrac{2}{5}\right)^{-3} = \left(\dfrac{5}{2}\right)^3 = \dfrac{5^3}{2^3} = \dfrac{125}{8}$

38) $\left(\dfrac{3}{5}\right)^{-2} = \left(\dfrac{5}{3}\right)^2 = \dfrac{5^2}{3^2} = \dfrac{25}{9}$

39) $2y^{-3} = \dfrac{2}{y^3}$

40) $13y^{-5} = \dfrac{13}{y^5}$

Effortless Math Education

EffortlessMath.com

41) $-20x^{-2} = -\frac{20}{x^2}$

42) $15a^{-2}b^3 = \frac{15b^3}{a^2}$

43) $23a^2b^{-4}c^{-8} = \frac{23a^2}{b^4c^8}$

44) $-4x^4y^{-2}2^{-7} = -\frac{4x^4}{y^2}$

45) $\frac{16y}{x^3y^{-4}} \to \frac{y}{y^{-4}} = y^{1-(-4)} = y^5 \to \frac{16y}{x^3y^{-4}} = \frac{16y^5}{x^3}$

46) $\frac{30a^{-3}b}{-100c^{-2}} \to -\frac{30\div 10}{100\div 10} = -\frac{3}{10} \to \frac{30a^{-3}b}{-100c^{-2}} = -\frac{3c^2b}{10a^3}$

47) $0.00518 = 5.18 \times 10^{-3}$

48) $0.000042 = 4.2 \times 10^{-5}$

49) $78,000 = 7.8 \times 10^4$

50) $92,000,000 = 9.2 \times 10^7$

51) $\sqrt{5} \times \sqrt{5} = \sqrt{25} = 5$

52) $\sqrt{25} - \sqrt{4} = 5 - 2 = 3$

53) $\sqrt{81} + \sqrt{36} = 9 + 6 = 15$

54) $\sqrt{4} \times \sqrt{25} = 2 + 5 = 10$

55) $\sqrt{2} \times \sqrt{18} = \sqrt{36} = 6$

56) $4\sqrt{2} + 3\sqrt{2} = 7\sqrt{2}$

57) $5\sqrt{7} + 2\sqrt{7} = 7\sqrt{7}$

58) $\sqrt{45} + 2\sqrt{5} = \sqrt{5 \times 9} + 2\sqrt{5} = 3\sqrt{5} + 2\sqrt{5} = 5\sqrt{5}$

Effortless Math Education

DAY 5: Expressions and Variables

Math topics that you'll learn in this chapter:

1. Simplifying Variable Expressions
2. Simplifying Polynomial Expressions
3. The Distributive Property
4. Evaluating One Variable
5. Evaluating Two Variables

Simplifying Variable Expressions

✰ In algebra, a variable is a letter used to stand for a number. The most common letters are $x, y, z, a, b, c, m,$ and n.

✰ An algebraic expression is an expression that contains integers, variables, and math operations such as addition, subtraction, multiplication, division, etc.

✰ In an expression, we can combine "like" terms. (values with same variable and same power)

Examples:

Example 1. Simplify. $(3x + 9x + 2) =$

Solution: In this expression, there are three terms: $3x$, $9x$, and 2. Two terms are "like terms": $3x$ and $9x$. Combine like terms. $3x + 9x = 12x$. Then: $(3x + 9x + 2) = 12x + 2$ (*remember you cannot combine variables and numbers.*)

Example 2. Simplify. $-17x^2 + 6x + 15x^2 - 13 =$

Solution: Combine "like" terms: $-17x^2 + 15x^2 = -2x^2$
Then: $-17x^2 + 6x + 15x^2 - 13 = -2x^2 + 6x - 13$

Example 3. Simplify. $5x - 18 - 6x^2 + 3x^2 =$

Solution: Combine like terms. Then:
$5x - 18 - 6x^2 + 3x^2 = -3x^2 + 5x - 18$

Example 4. Simplify. $-5x - 4x^2 + 9x - 11x^2 =$

Solution: Combine "like" terms: $-5x + 9x = 4x$, and $-4x^2 - 11x^2 = -15x^2$
Then: $-5x - 4x^2 + 9x - 11x^2 = 4x - 15x^2$. Write in standard form (biggest powers first): $4x - 15x^2 = -15x^2 + 4x$

EffortlessMath.com

Simplifying Polynomial Expressions

☆ In mathematics, a polynomial is an expression consisting of variables and coefficients that involves only the operations of addition, subtraction, multiplication, and non–negative integer exponents of variables.

$$P(x) = a_n x^n + a_{n-1} x^{n-1} + \ldots + a_2 x^2 + a_1 x + a_0$$

☆ Polynomials must always be simplified as much as possible. It means you must add together any like terms. (values with same variable and same power)

Examples:

Example 1. Simplify this Polynomial Expressions. $-2x^2 + 9x^3 + 5x^3 - 7x^4$

Solution: Combine "like" terms: $9x^3 + 5x^3 = 14x^3$

Then: $-2x^2 + 9x^3 + 5x^3 - 7x^4 = -2x^2 + 14x^3 - 7x^4$
Now, write the expression in standard form:
$-2x^2 + 14x^3 - 7x^4 = -7x^4 + 14x^3 - 2x^2$

Example 2. Simplify this expression. $(4x^2 - x^3) - (-6x^3 + 3x^2) =$

Solution: First, multiply $(-)$ into $(-6x^3 + 3x^2)$:

$(4x^2 - x^3) - (-6x^3 + 3x^2) = 4x^2 - x^3 + 6x^3 - 3x^2$
Then combine "like" terms: $4x^2 - x^3 + 6x^3 - 3x^2 = x^2 + 5x^3$
And write in standard form: $x^2 + 5x^3 = 5x^3 + x^2$

Example 3. Simplify. $-2x^3 + 6x^4 - 5x^2 - 14x^4 =$

Solution: Combine "like" terms: $6x^4 - 14x^4 = -8x^4$
Then: $-2x^3 + 6x^4 - 5x^2 - 14x^4 = -2x^3 - 8x^4 - 5x^2$
And write in standard form: $-2x^3 - 8x^4 - 5x^2 = -8x^4 - 2x^3 - 5x^2$

The Distributive Property

✯ The distributive property (or the distributive property of multiplication over addition and subtraction) simplifies and solves expressions in the form of: $a(b + c)$ or $a(b - c)$

✯ The distributive property is multiplying a term outside the parentheses by the terms inside.

✯ Distributive Property rule: $a(b + c) = ab + ac$

Examples:

Example 1. Simply using the distributive property. $(3)(4x - 9)$

Solution: Use Distributive Property rule: $a(b + c) = ab + ac$

$(3)(4x - 9) = (3 \times 4x) + (3) \times (-9) = 12x - 27$

Example 2. Simply. $(-4)(-3x + 8)$

Solution: Use Distributive Property rule: $a(b + c) = ab + ac$

$(-4)(-3x + 8) = (-4 \times (-3x)) + (-4) \times (8) = 12x - 32$

Example 3. Simply. $(5)(3x + 4) - 13x$

Solution: First, simplify $(5)(3x + 4)$ using the distributive property.

Then: $(5)(3x + 4) = 15x + 20$

Now combine like terms: $(5)(3x + 4) - 13x = 15x + 20 - 13x$

In this expression, $15x$ and $-13x$ are "like terms" and we can combine them.

$15x - 13x = 2x$. Then: $15x + 20 - 13x = 2x + 20$

Evaluating One Variable

☆ To evaluate one variable expressions, find the variable and substitute a number for that variable.

☆ Perform the arithmetic operations.

Examples:

Example 1. Calculate this expression for $x = 1$. $9 + 8x$

Solution: First, substitute 1 for x.
Then: $9 + 8x = 9 + 8(1)$
Now, use order of operation to find the answer: $9 + 8(1) = 9 + 8 = 17$

Example 2. Evaluate this expression for $x = -2$. $7x - 3$

Solution: First, substitute -2 for x.
Then: $7x - 3 = 7(-2) - 3$
Now, use order of operation to find the answer: $7(-2) - 3 = -14 - 3 = -17$

Example 3. Find the value of this expression when $x = 3$. $(12 - 2x)$

Solution: First, substitute 3 for x,
Then: $12 - 2x = 12 - 2(3) = 12 - 6 = 6$

Example 4. Solve this expression for $x = -4$. $11 + 5x$

Solution: Substitute -4 for x.
Then: $11 + 5x = 11 + 5(-4) = 11 - 20 = -9$

Evaluating Two Variables

✩ To evaluate an algebraic expression, substitute a number for each variable.

✩ Perform the arithmetic operations to find the value of the expression.

Examples:

Example 1. Calculate this expression for $a = -2$ and $b = 3$. $(2a - 6b)$

Solution: First, substitute -2 for a, and 3 for b.
Then: $2a - 6b = 2(-2) - 6(3)$
Now, use order of operation to find the answer: $2(-2) - 6(3) = -4 - 18 = -22$

Example 2. Evaluate this expression for $x = -3$ and $y = 4$. $(2x - 4y)$

Solution: Substitute -3 for x, and 4 for y.
Then: $2x - 4y = 2(-3) - 4(4) = -6 - 16 = -22$

Example 3. Find the value of this expression $3(-4a + 2b)$, when $a = -2$ and $b = -3$.

Solution: Substitute -2 for a, and -3 for b.
Then: $3(-4a + 2b) = 3(-4(-2) + 2(-3)) = 3(8 - 6) = 3(2) = 6$

Example 4. Evaluate this expression. $-5x - 3y$, $x = 2$, $y = -6$

Solution: Substitute 2 for x, and -6 for y and simplify.
Then: $-5x - 3y = -5(2) - 3(-6) = -10 + 18 = 8$

Day 5: Practices

✎ **Simplify each expression.**

1) $2 - 3x - 1 =$

2) $-6 - 2x + 8 =$

3) $11x - 6x - 4 =$

4) $-16x + 25x - 5 =$

5) $5x + 5 - 15x =$

6) $4 + 5x - 6x - 5 =$

7) $3x + 10 - 2x - 20 =$

8) $-3 - 2x^2 - 5 + 3x =$

9) $-7 + 9x^2 - 2 + 2x =$

10) $4x^2 + 2x - 12x - 5 =$

11) $2x^2 - 3x - 5x + 6 - 9 =$

12) $x^2 - 6x - x + 2 - 3 =$

13) $10x^2 - x - 8x + 3 - 10 =$

14) $4x^2 - 7x - x^2 + 2x + 5 =$

✎ **Simplify each polynomial.**

15) $4x^2 + 3x^3 - x^2 + x =$

16) $5x^4 + x^5 - x^4 + 4x^2 =$

17) $15x^3 + 12x - 6x^2 - 9x^3 =$

18) $(7x^3 - 2x^2) + (6x^2 - 13x) =$

19) $(9x^4 + 6x^3) + (11x^3 - 5x^4) =$

20) $(15x^5 - 5x^3) - (4x^3 + 6x^2) =$

21) $(15x^4 + 7x^3) - (3x^3 - 26) =$

22) $(22x^4 + 6x^3) - (-2x^3 - 4x^4) =$

23) $(x^2 + 6x^3) + (-19x^2 + 6x^3) =$

24) $(2x^4 - x^3) + (-5x^3 - 7x^4) =$

Use the distributive property to simply each expression.

25) $3(5 + x) =$

26) $5(4 - x) =$

27) $6(2 - 5x) =$

28) $(4 - 3x)7 =$

29) $8(3 - 3x) =$

30) $(-1)(-6 + 2x) =$

31) $(-5)(3x - 3) =$

32) $(-x + 10)(-3)$

33) $(-2)(2 - 6x) =$

34) $(-6x - 4)(-7) =$

Evaluate each expression using the value given.

35) $x = 3 \rightarrow 12 - x =$

36) $x = 5 \rightarrow x + 7 =$

37) $x = 3 \rightarrow 3x - 5 =$

38) $x = 2 \rightarrow 18 - 3x =$

39) $x = 7 \rightarrow 5x - 4 =$

40) $x = 6 \rightarrow 21 - x =$

41) $x = 5 \rightarrow 10x - 20 =$

42) $x = -5 \rightarrow 4 - x =$

43) $x = -2 \rightarrow 25 - 3x =$

44) $x = -7 \rightarrow 16 - x =$

45) $x = -13 \rightarrow 40 - 2x =$

46) $x = -4 \rightarrow 20x - 6 =$

47) $x = -6 \rightarrow -11x - 19 =$

48) $x = -8 \rightarrow -1 - 3x =$

Evaluate each expression using the values given.

49) $x = 3, y = 2 \rightarrow 3x + 2y =$

50) $a = 4, b = 1 \rightarrow 2a - 6b =$

51) $x = 5, y = 7 \rightarrow 2x - 4y - 5 =$

52) $a = -3, b = 4 \rightarrow -3a + 4b + 2 =$

53) $x = -4, y = -3 \rightarrow 2x - 6 - 4y =$

Day 5: Answers

1) $2 - 3x - 1 \rightarrow 2 - 1 = 1 \rightarrow 2 - 3x - 1 = -3x + 1$

2) $-6 - 2x + 8 \rightarrow -6 + 8 = 2 \rightarrow -6 - 2x + 8 = -2x + 2$

3) $11x - 6x - 4 \rightarrow 11x - 6x = 5x \rightarrow 11x - 6x - 4 = 5x - 4$

4) $-16x + 25x - 5 \rightarrow -16x + 25x = 9x \rightarrow -16x + 25x - 5 = 9x - 5$

5) $5x + 5 - 15x \rightarrow 5x - 15x = -10x \rightarrow 5x + 5 - 15x = -10x + 5$

6) $4 + 5x - 6x - 5 \rightarrow 4 - 5 = -1, 5x - 6x = -x \rightarrow 4 + 5x - 6x - 5 = -x - 1$

7) $3x + 10 - 2x - 20 \rightarrow 10 - 20 = -10, 3x - 2x = x \rightarrow 3x + 10 - 2x - 20 = x - 10$

8) $-3 - 2x^2 - 5 + 3x \rightarrow -3 - 5 = -8 \rightarrow -3 - 2x^2 - 5 + 3x = -2x^2 + 3x - 8$

9) $-7 + 9x^2 - 2 + 2x \rightarrow -7 - 2 = -9 \rightarrow -7 + 9x^2 - 2 + 2x = 9x^2 + 2x - 9$

10) $4x^2 + 2x - 12x - 5 \rightarrow 2x - 12x = -10x \rightarrow 4x^2 + 2x - 12x - 5 = 4x^2 - 10x - 5$

11) $2x^2 - 3x - 5x + 6 - 9 \rightarrow -3x - 5x = -8x, 6 - 9 = -3 \rightarrow$
    $2x^2 - 3x - 5x + 6 - 9 = 2x^2 - 8x - 3$

12) $x^2 - 6x - x + 2 - 3 \rightarrow -6x - x = -7x, 2 - 3 = -1 \rightarrow x^2 - 6x - x + 2 - 3 = x^2 - 7x - 1$

13) $10x^2 - x - 8x + 3 - 10 \rightarrow -x - 8x = -9x, 3 - 10 = -7 \rightarrow 10x^2 - x - 8x + 3 - 10 = 10x^2 - 9x - 7$

14) $4x^2 - 7x - x^2 + 2x + 5 \rightarrow 4x^2 - x^2 = 3x^2, -7x + 2x = -5x \rightarrow$
    $4x^2 - 7x - x^2 + 2x + 5 = 3x^2 - 5x + 5$

15) $4x^2 + 3x^3 - x^2 + x \rightarrow 4x^2 - x^2 = 3x^2 \rightarrow 4x^2 + 3x^3 - x^2 + x = 3x^3 + 3x^2 + x$

16) $5x^4 + x^5 - x^4 + 4x^2 \rightarrow 5x^4 - x^4 = 4x^4 \rightarrow 5x^4 + x^5 - x^4 + 4x^2 =$
    $4x^4 + x^5 + 4x^2 = x^5 + 4x^4 + 4x^2$

17) $15x^3 + 12x - 6x^2 - 9x^3 \to 15x^3 - 9x^3 = 6x^3 \to 15x^3 + 12x - 6x^2 - 9x^3 =$
$6x^3 + 12x - 6x^2 = 6x^3 - 6x^2 + 12x$

18) $(7x^3 - 2x^2) + (6x^2 - 13x) = 7x^3 - 2x^2 + 6x^2 - 13x \to -2x^2 + 6x^2 = 4x^2 \to$
$7x^3 - 2x^2 + 6x^2 - 13x = 7x^3 + 4x^2 - 13x$

19) $(9x^4 + 6x^3) + (11x^3 - 5x^4) = 9x^4 + 6x^3 + 11x^3 - 5x^4 \to 9x^4 - 5x^4 = 4x^4,$
$6x^3 + 11x^3 = 17x^3 \to 9x^4 + 6x^3 + 11x^3 - 5x^4 = 4x^4 + 17x^3$

20) $(15x^5 - 5x^3) - (4x^3 + 6x^2) = 15x^5 - 5x^3 - 4x^3 - 6x^2 \to -5x^3 - 4x^3 = -9x^3 \to$
$15x^5 - 5x^3 - 4x^3 - 6x^2 = 15x^5 - 9x^3 - 6x^2$

21) $(15x^4 + 7x^3) - (3x^3 - 26) = 15x^4 + 7x^3 - 3x^3 + 26 \to 7x^3 - 3x^3 = 4x^3 \to$
$15x^4 + 7x^3 - 3x^3 + 26 = 15x^4 + 4x^3 + 26$

22) $(22x^4 + 6x^3) - (-2x^3 - 4x^4) = 22x^4 + 6x^3 + 2x^3 + 4x^4 \to 22x^4 + 4x^4 =$
$26x^4, 6x^3 + 2x^3 = 8x^3 \to 22x^4 + 6x^3 + 2x^3 + 4x^4 = 26x^4 + 8x^3$

23) $(x^2 + 6x^3) + (-19x^2 + 6x^3) = x^2 + 6x^3 - 19x^2 + 6x^3 \to 6x^3 + 6x^3 = 12x^3,$
$-19x^2 + x^2 = -18x^2 \to x^2 + 6x^3 - 19x^2 + 6x^3 = 12x^3 - 18x^2$

24) $(2x^4 - x^3) + (-5x^3 - 7x^4) = 2x^4 - x^3 - 5x^3 - 7x^4 \to 2x^4 - 7x^4 =$
$-5x^4, -x^3 - 5x^3 = -6x^3 \to 2x^4 - x^3 - 5x^3 - 7x^4 = -5x^4 - 6x^3$

25) $3(5 + x) = (3) \times (5) + (3) \times x = 15 + 3x = 3x + 15$

26) $5(4 - x) = (5) \times (4) + (5) \times (-x) = 20 + (-5x) = -5x + 20$

27) $6(2 - 5x) = (6) \times (2) + (6) \times (-5x) = 12 + (-30x) = -30x + 12$

28) $(4 - 3x)7 = (4) \times (7) + (-3x) \times (7) = 28 + (-21x) = -21x + 28$

29) $8(3 - 3x) = (8) \times (3) + (8) \times (-3x) = 24 + (-24x) = -24x + 24$

30) $(-1)(-6 + 2x) = (-1) \times (-6) + (-1) \times (2x) = 6 + (-2x) = -2x + 6$

31) $(-5)(3x - 3) = (-5) \times (3x) + (-5) \times (-3) = -15x + 15$

32) $(-x + 10)(-3) = (-x) \times (-3) + (10) \times (-3) = 3x - 30$

33) $(-2)(2 - 6x) = (-2) \times (2) + (-2) \times (-6x) = -4 + 12x = 12x - 4$

34) $(-6x - 4)(-7) = (-6x) \times (-7) + (-4) \times (-7) = 42x + 28$

35) $x = 3 \to 12 - x = 12 - 3 = 9$

36) $x = 5 \to x + 7 = 5 + 7 = 12$

37) $x = 3 \to 3x - 5 = (3) \times (3) - 5 = 9 - 5 = 4$

38) $x = 2 \to 18 - 3x = 18 - (3) \times (2) = 18 - 6 = 12$

39) $x = 7 \to 5x - 4 = (5) \times (7) - 4 = 35 - 4 = 31$

40) $x = 6 \to 21 - x = 21 - 6 = 15$

41) $x = 5 \to 10x - 20 = (10) \times (5) - 20 = 50 - 20 = 30$

42) $x = -5 \to 4 - x = 4 - (-5) = 4 + 5 = 9$

43) $x = -2 \to 25 - 3x = 25 - (3) \times (-2) = 25 - (-6) = 25 + 6 = 31$

44) $x = -7 \to 16 - x = 16 - (-7) = 16 + 7 = 23$

45) $x = -13 \to 40 - 2x = 40 - (2) \times (-13) = 40 - (-26) = 40 + 26 = 66$

46) $x = -4 \to 20x - 6 = 20 \times (-4) - 6 = -80 - 6 = -86$

47) $x = -6 \to -11x - 19 = (-11) \times (-6) - 19 = 66 - 19 = 47$

48) $x = -8 \to -1 - 3x = (-1) - (3) \times (-8) = -1 - (-24) = -1 + 24 = 23$

49) $x = 3, y = 2 \to 3x + 2y = 3 \times (3) + 2(2) = 9 + 4 = 13$

50) $a = 4, b = 1 \to 2a - 6b = 2 \times (4) - 6(1) = 8 - 6 = 2$

51) $x = 5, y = 7 \to 2x - 4y - 5 = 2 \times (5) - 4(7) - 5 = 10 - 28 - 5 = -23$

52) $a = -3, b = 4 \to -3a + 4b + 2 = -3 \times (-3) + 4(4) + 2 = 9 + 16 + 2 = 27$

53) $x = -4, y = -3 \to 2x - 6 - 4y = 2 \times (-4) - 6 - 4(-3) = -8 - 6 + 12 = -2$

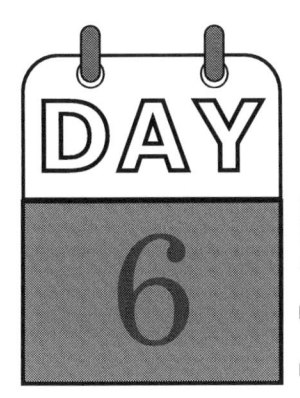

DAY 6 Equations and Inequalities

Math topics that you'll learn in this chapter:

1. One-Step Equations
2. Multi-Step Equations
3. System of Equations
4. Graphing Single–Variable Inequalities
5. One-Step Inequalities
6. Multi-Step Inequalities

One–Step Equations

☆ The values of two expressions on both sides of an equation are equal. Example: $ax = b$. In this equation, ax is equal to b.

☆ Solving an equation means finding the value of the variable.

☆ You only need to perform one Math operation to solve the one-step equations.

☆ To solve a one-step equation, find the inverse (opposite) operation is being performed.

☆ The inverse operations are:

- Addition and subtraction
- Multiplication and division

Examples:

Example 1. Solve this equation for x. $6x = 18 \rightarrow x = ?$

Solution: Here, the operation is multiplication (variable x is multiplied by 6) and its inverse operation is division. To solve this equation, divide both sides of equation by 6: $6x = 18 \rightarrow \frac{6x}{6} = \frac{18}{6} \rightarrow x = 3$

Example 2. Solve this equation. $x + 5 = 0 \rightarrow x = ?$

Solution: In this equation, 5 is added to the variable x. The inverse operation of addition is subtraction. To solve this equation, subtract 5 from both sides of the equation: $x + 5 - 5 = 0 - 5$. Then: $x + 5 - 5 = 0 - 5 \rightarrow x = -5$

Example 3. Solve this equation for x. $x - 11 = 0$

Solution: Here, the operation is subtraction and its inverse operation is addition. To solve this equation, add 11 to both sides of the equation:

$$x - 11 + 11 = 0 + 11 \rightarrow x = 11$$

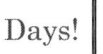

Multi–Step Equations

☆ To solve a multi-step equation, combine "like" terms on one side.

☆ Bring variables to one side by adding or subtracting.

☆ Simplify using the inverse of addition or subtraction.

☆ Simplify further by using the inverse of multiplication or division.

☆ Check your solution by plugging the value of the variable into the original equation.

Examples:

Example 1. Solve this equation for x. $5x - 6 = 26 - 3x$

Solution: First, bring variables to one side by adding $3x$ to both sides. Then:
$5x - 6 + 3x = 26 - 3x + 3x \rightarrow 5x - 6 + 3x = 26$.
Simplify: $8x - 6 = 26$. Now, add 6 to both sides of the equation:
$8x - 6 + 6 = 26 + 6 \rightarrow 8x = 32 \rightarrow$ Divide both sides by 8:
$8x = 32 \rightarrow \frac{8x}{8} = \frac{32}{8} \rightarrow x = 4$

Let's check this solution by substituting the value of 4 for x in the original equation:
$x = 4 \rightarrow 5x - 6 = 26 - 3x \rightarrow 5(4) - 6 = 26 - 3(4) \rightarrow 20 - 6 = 26 - 12 \rightarrow 14 = 14$
The answer $x = 4$ is correct.

Example 2. Solve this equation for x. $6x - 3 = 15$

Solution: Add 3 to both sides of the equation.
$6x - 3 = 15 \rightarrow 6x - 3 + 3 = 15 + 3 \rightarrow 6x = 18$
Divide both sides by 6, then: $6x = 18 \rightarrow \frac{6x}{6} = \frac{18}{6} \rightarrow x = 3$

Now, check the solution:
$x = 4 \rightarrow 6(3) - 3 = 15 \rightarrow 18 - 3 = 15 \rightarrow 15 = 15$
The answer $x = 4$ is correct.

System of Equations

☆ A system of equations contains two equations and two variables. For example, consider the system of equations: $x - y = 1$ and $x + y = 5$

☆ The easiest way to solve a system of equations is using the elimination method. The elimination method uses the addition property of equality. You can add the same value to each side of an equation.

☆ For the first equation above, you can add $x + y$ to the left side and 5 to the right side of the first equation: $x - y + (x + y) = 1 + 5$. Now, if you simplify, you get: $x - y + (x + y) = 1 + 5 \to 2x = 6 \to x = 3$. Now, substitute 3 for the x in the first equation: $3 - y = 1$. By solving this equation, $y = 2$

Example:

What is the value of $x + y$ in this system of equations?

$$\begin{cases} -x + y = 18 \\ 2x - 6y = -12 \end{cases}$$

Solution: Solving the system of equations by elimination:
Multiply the first equation by (2), then add it to the second equation.

$\begin{array}{l} 2(-x + y = 18) \\ 2x - 6y = -12 \end{array} \Rightarrow \begin{array}{l} -2x + 2y = 36 \\ 2x - 6y = -12 \end{array} \Rightarrow (-2x) + 2x + 2y - 6y = 36 - 12 \Rightarrow -4y = 24 \Rightarrow$
$y = -6$

Plug in the value of y into one of the equations and solve for x.
$-x + (-6) = 18 \Rightarrow -x - 6 = 18 \Rightarrow -x = 24 \Rightarrow x = -24$
Thus, $x + y = -24 - 6 = -30$

Graphing Single–Variable Inequalities

☆ An inequality compares two expressions using an inequality sign.

☆ Inequality signs are: "less than" <, "greater than" >, "less than or equal to" ≤, and "greater than or equal to" ≥.

☆ To graph a single–variable inequality, find the value of the inequality on the number line.

☆ For less than (<) or greater than (>) draw an open circle on the value of the variable. If there is an equal sign too, then use a filled circle.

☆ Draw an arrow to the right for greater or to the left for less than.

Examples:

Example 1. Draw a graph for this inequality. $x > 3$

Solution: Since the variable is greater than 3, then we need to find 3 in the number line and draw an open circle on it. Then, draw an arrow to the right.

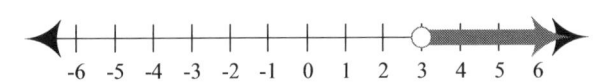

Example 2. Graph this inequality. $x \leq -1$.

Solution: Since the variable is less than or equal to -1, then we need to find -1 on the number line and draw a filled circle on it. Then, draw an arrow to the left.

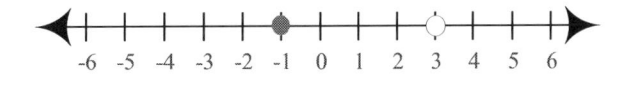

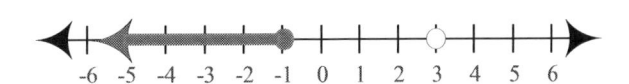

One–Step Inequalities

☆ An inequality compares two expressions using an inequality sign.

☆ Inequality signs are: "less than" <, "greater than" >, "less than or equal to" ≤, and "greater than or equal to" ≥.

☆ You only need to perform one Math operation to solve the one-step inequalities.

☆ To solve one-step inequalities, find the inverse (opposite) operation is being performed.

☆ For dividing or multiplying both sides by negative numbers, flip the direction of the inequality sign.

Examples:

Example 1. Solve this inequality for x. $x + 7 \geq 2$

Solution: The inverse (opposite) operation of addition is subtraction. In this inequality, 7 is added to x. To isolate x we need to subtract 7 from both sides of the inequality.
Then: $x + 7 \geq 2 \rightarrow x + 7 - 7 \geq 2 - 7 \rightarrow x \geq -5$. The solution is: $x \geq -5$

Example 2. Solve the inequality. $x - 2 > -12$

Solution: 2 is subtracted from x. Add 2 to both sides.
$x - 2 > -12 \rightarrow x - 2 + 2 > -12 + 2 \rightarrow x > -10$

Example 3. Solve. $6x \leq -36$

Solution: 6 is multiplied to x. Divide both sides by 6.
Then: $6x \leq -36 \rightarrow \frac{6x}{6} \leq \frac{-36}{6} \rightarrow x \leq -6$

Example 4. Solve. $-2x \leq 10$

Solution: -2 is multiplied to x. Divide both sides by -2. Remember when dividing or multiplying both sides of an inequality by negative numbers, flip the direction of the inequality sign.
Then: $-2x \leq 10 \rightarrow \frac{-2x}{-2} \geq \frac{10}{-2} \rightarrow x \geq -5$

Multi–Step Inequalities

★ To solve a multi-step inequality, combine "like" terms on one side.

★ Bring variables to one side by adding or subtracting.

★ Isolate the variable.

★ Simplify using the inverse of addition or subtraction.

★ Simplify further by using the inverse of multiplication or division.

★ For dividing or multiplying both sides by negative numbers, flip the direction of the inequality sign.

Examples:

Example 1. Solve this inequality. $4x - 1 \leq 23$

Solution: In this inequality, 1 is subtracted from $4x$. The inverse of subtraction is addition. Add 1 to both sides of the inequality:
$4x - 1 + 1 \leq 23 + 1 \rightarrow 4x \leq 24$
Now, divide both sides by 4. Then: $4x \leq 24 \rightarrow \frac{4x}{4} \leq \frac{24}{4} \rightarrow x \leq 6$
The solution of this inequality is $x \leq 6$.

Example 2. Solve this inequality. $2x - 6 < 18$

Solution: First, add 6 to both sides: $2x - 6 + 6 < 18 + 6$
Then simplify: $2x - 6 + 6 < 18 + 6 \rightarrow 2x < 24$
Now divide both sides by 2: $\frac{2x}{2} < \frac{24}{2} \rightarrow x < 12$

Example 3. Solve this inequality. $-4x - 8 \geq 12$

Solution: First, add 8 to both sides:
$-4x - 8 + 8 \geq 12 + 8 \rightarrow -4x \geq 20$
Divide both sides by -4. Remember that you need to flip the direction of inequality sign. $-4x \geq 20 \rightarrow \frac{-4x}{-4} \leq \frac{20}{-4} \rightarrow x \leq -5$

Day 6: Practices

✎ **Solve each equation. (One−Step Equations)**

1) $x + 2 = 5 \rightarrow x =$

2) $8 = 13 + x \rightarrow x =$

3) $-6 = 7 + x \rightarrow x =$

4) $x - 5 = -3 \rightarrow x =$

5) $-13 = x - 15 \rightarrow x =$

6) $-10 + x = -4 \rightarrow x =$

7) $-19 + x = 7 \rightarrow x =$

8) $-6x = 24 \rightarrow x =$

9) $\frac{x}{4} = -5 \rightarrow x =$

10) $-2x = -4 \rightarrow x =$

✎ **Solve each equation. (Multi−Step Equations)**

11) $2(x + 5) = 16 \rightarrow x =$

12) $-6(3 - x) = 18 \rightarrow x =$

13) $25 = -5(x + 4) \rightarrow x =$

14) $-12 = 6(9 + x) \rightarrow x =$

15) $11(x + 5) = -22 \rightarrow x =$

16) $-27 - 36x = 45 \rightarrow x =$

17) $3x - 4 = x - 12 \rightarrow x =$

18) $-8x + x - 11 = 24 \rightarrow x =$

✎ **Solve each system of equations.**

19) $\begin{cases} x + 4y = 29 \\ x + 2y = 5 \end{cases}$ $x = \underline{}$ $y = \underline{}$

20) $\begin{cases} 2x + y = 36 \\ x + 4y = 4 \end{cases}$ $x = \underline{}$ $y = \underline{}$

21) $\begin{cases} 2x + 5y = 15 \\ x + y = 6 \end{cases}$ $x = \underline{}$ $y = \underline{}$

22) $\begin{cases} 2x - 2y = -16 \\ -9x + 2y = -19 \end{cases}$ $x = \underline{}$ $y = \underline{}$

Equations and Inequalities
Day 6: Practices

✎ **Draw a graph for each inequality.**

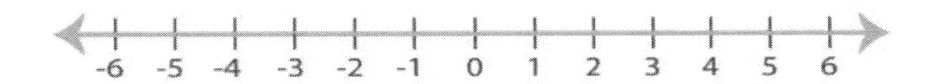

23) $x \leq 1$

24) $x > -4$

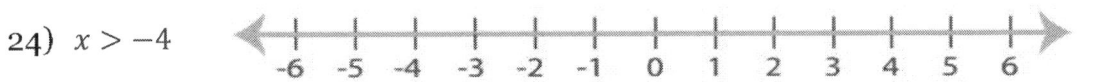

✎ **Solve each inequality and graph it.**

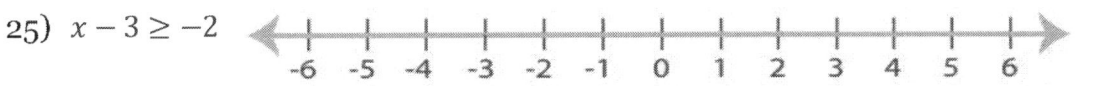

25) $x - 3 \geq -2$

26) $7x - 6 < 8$

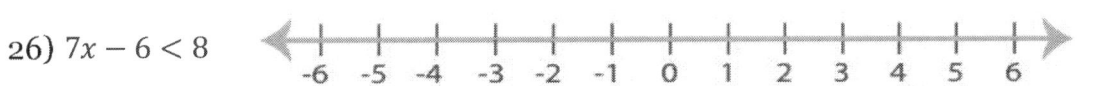

✎ **Solve each inequality.**

27) $x + 11 > 3$

28) $x + 4 > 1$

29) $-6 + 3x \leq 21$

30) $-5 + 4x \leq 19$

31) $4 + 9x \leq 31$

32) $8(x + 3) \geq -16$

33) $3(6 + x) \geq 18$

34) $3(x - 2) < -9$

35) $15 + 9x < -30$

36) $3(6 - x) \geq -27$

37) $4(x - 5) \geq -32$

38) $6(x + 4) < -24$

39) $7(x - 8) \geq -49$

40) $-(-6 - 5x) > -39$

41) $2(1 - 2x) > -66$

42) $-3(3 - 2x) > -33$

Day 6: Answers

1) $x + 2 = 5 \to x = 5 - 2 = 3$

2) $8 = 13 + x \to x = 8 - 13 = -5$

3) $-6 = 7 + x \to x = -6 - 7 = -13$

4) $x - 5 = -3 \to x = -3 + 5 = 2$

5) $-13 = x - 15 \to x = -13 + 15 = 2$

6) $-10 + x = -4 \to x = -4 + 10 = 6$

7) $-19 + x = 7 \to x = 7 + 19 = 26$

8) $-6x = 24 \to x = \frac{24}{-6} = -4$

9) $\frac{x}{4} = -5 \to x = -5 \times 4 = -20$

10) $-2x = -4 \to x = \frac{-4}{-2} = 2$

11) $2(x + 5) = 16 \to \frac{2(x+5)}{2} = \frac{16}{2} \to (x + 5) = 8 \to x = 8 - 5 = 3$

12) $-6(3 - x) = 18 \to \frac{-6(3-x)}{-6} = \frac{18}{-6} \to (3 - x) = -3 \to x = 3 + 3 = 6$

13) $25 = -5(x + 4) \to \frac{25}{-5} = \frac{-5(x+4)}{-5} \to -5 = (x + 4) \to x = -5 - 4 = -9$

14) $-12 = 6(9 + x) \to \frac{-12}{6} = \frac{6(9+x)}{6} \to -2 = (9 + x) \to x = -2 - 9 = -11$

15) $11(x + 5) = -22 \to \frac{11(x+5)}{11} = \frac{-22}{11} \to (x + 5) = -2 \to x = -2 - 5 = -7$

16) $-27 - 36x = 45 \to -36x = 45 + 27 \to -36x = 72 \to \frac{-36x}{-2} = \frac{72}{-2} \to x = -2$

17) $3x - 4 = x - 12 \to 3x - 4 - x = x - 12 - x \to 2x - 4 = -12 \to 2x = -12 + 4 \to$
    $2x = -8 \to \frac{2x}{2} = \frac{-8}{2} \to x = -4$

18) $-8x + x - 11 = 24 \to -7x = 24 + 11 \to \frac{-7x}{-7} = \frac{35}{-7} \to x = -5$

19) $\begin{cases} x + 4y = 29 \\ x + 2y = 5 \end{cases} \to \begin{array}{c} -(x + 4y = 29) \\ x + 2y = 5 \end{array} \to \begin{array}{c} -x - 4y = -29 \\ x + 2y = 5 \end{array} \to (-x) + x - 4y + 2y =$
    $-29 + 5 \to -2y = -24 \to \frac{-2y}{-2} = \frac{-24}{-2} \to y = 12$

Plug in the value of y into one of the equations and solve for x.
$x + 2y = 5 \to x + 2(12) = 5 \to x + 24 = 5 \to x = 5 - 24 = -19$

20) $\begin{cases} 2x+y=36 \\ x+4y=4 \end{cases} \rightarrow \begin{matrix} 2x+y=36 \\ -2(x+4y=4) \end{matrix} \rightarrow \begin{matrix} 2x+y=36 \\ -2x-8y=-8 \end{matrix} \rightarrow 2x-2x+y-8y=36-8 \rightarrow -7y=28 \rightarrow \frac{-7y}{-7}=\frac{28}{-7} \rightarrow y=-4$

Plug in the value of y into one of the equations and solve for x.

$x+4y=4 \rightarrow x+4(-4)=4 \rightarrow x-16=4 \rightarrow x=4+16=20$

21) $\begin{cases} 2x+5y=15 \\ x+y=6 \end{cases} \rightarrow \begin{matrix} 2x+5y=15 \\ -2(x+y=6) \end{matrix} \rightarrow \begin{matrix} 2x+5y=15 \\ -2x-2y=-12 \end{matrix} \rightarrow 2x-2x+5y-2y=15-12 \rightarrow 3y=3 \rightarrow \frac{3y}{3}=\frac{3}{3} \rightarrow y=1$

Plug in the value of y into one of the equations and solve for x.

$x+y=6 \rightarrow x+1=6 \rightarrow x=6-1=5$

22) $\begin{cases} 2x-2y=-16 \\ -9x+2y=-19 \end{cases} \rightarrow 2x-9x-2y+2y=-16-19 \rightarrow -7x=-35 \rightarrow \frac{-7x}{-7}=\frac{-35}{-7} \rightarrow x=5$

Plug in the value of x into one of the equations and solve for y.

$2x-2y=-16 \rightarrow 2(5)-2y=-16 \rightarrow 10-2y=-16 \rightarrow 10+16=2y \rightarrow 26=2y \rightarrow \frac{26}{2}=\frac{2y}{2} \rightarrow y=13$

23) $x \leq 1$

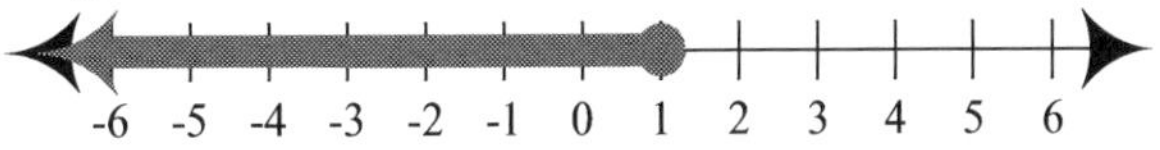

24) $x > -4$

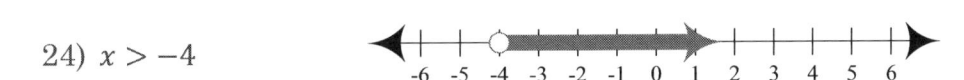

25) $x-3 \geq -2 \rightarrow x \geq -2+3 \rightarrow x \geq 1$

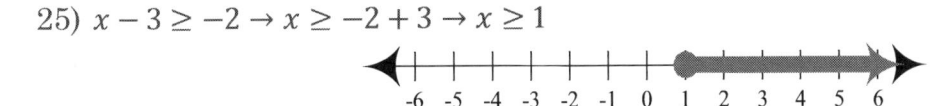

26) $7x-6<8 \rightarrow 7x<8+6 \rightarrow 7x<14 \rightarrow x<\frac{14 \div 7}{7 \div 7} \rightarrow x<2$

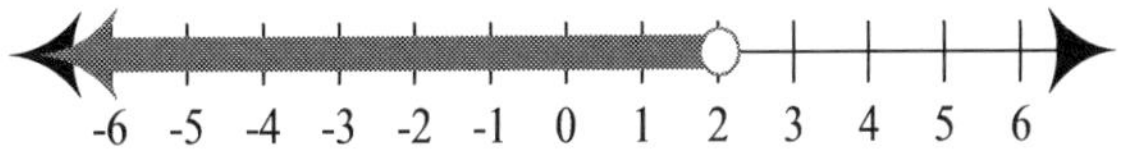

27) $x+11>3 \rightarrow x>3-11 \rightarrow x>-8$

28) $x+4>1 \rightarrow x>1-4 \rightarrow x>-3$

29) $-6+3x \leq 21 \rightarrow 3x \leq 21+6 \rightarrow 3x \leq 27 \rightarrow x \leq \frac{27 \div 3}{3 \div 3} \rightarrow x \leq 9$

30) $-5+4x \leq 19 \rightarrow 4x \leq 19+5 \rightarrow 4x \leq 24 \rightarrow x \leq \frac{24 \div 4}{4 \div 4} \rightarrow x \leq 6$

31) $4 + 9x \leq 31 \rightarrow 9x \leq 31 - 4 \rightarrow 9x \leq 27 \rightarrow x \leq \frac{27 \div 9}{9 \div 9} \rightarrow x \leq 3$

32) $8(x + 3) \geq -16 \rightarrow x + 3 \geq \frac{-16}{8} \rightarrow x + 3 \geq -2 \rightarrow x \geq -2 - 3 \rightarrow x \geq -5$

33) $3(6 + x) \geq 18 \rightarrow 6 + x \geq \frac{18}{3} \rightarrow 6 + x \geq 6 \rightarrow x \geq 6 - 6 \rightarrow x \geq 0$

34) $3(x - 2) < -9 \rightarrow x - 2 < \frac{-9}{3} \rightarrow x - 2 < -3 \rightarrow x < -3 + 2 \rightarrow x < -1$

35) $15 + 9x < -30 \rightarrow 9x < -30 - 15 \rightarrow 9x < -45 \rightarrow x < \frac{-45}{9} \rightarrow x < -5$

36) $3(6 - x) \geq -27 \rightarrow 6 - x \geq \frac{-27}{3} \rightarrow 6 - x \geq -9 \rightarrow -x \geq -9 - 6 \rightarrow \frac{-x}{-1} \leq \frac{-15}{-1} \rightarrow$
$x \leq 15$

37) $4(x - 5) \geq -32 \rightarrow x - 5 \geq \frac{-32}{4} \rightarrow x - 5 \geq -8 \rightarrow x \geq -8 + 5 \rightarrow x \geq -3$

38) $6(x + 4) < -24 \rightarrow x + 4 < \frac{-24}{6} \rightarrow x + 4 < -4 \rightarrow x < -4 - 4 \rightarrow x < -8$

39) $7(x - 8) \geq -49 \rightarrow x - 8 \geq \frac{-49}{7} \rightarrow x - 8 \geq -7 \rightarrow x \geq -7 + 8 \rightarrow x \geq 1$

40) $-(-6 - 5x) > -39 \rightarrow 6 + 5x > -39 \rightarrow 5x > -39 - 6 \rightarrow 5x > -45 \rightarrow$
$x > \frac{-45}{5} \rightarrow x > -9$

41) $2(1 - 2x) > -66 \rightarrow 1 - 2x > \frac{-66}{2} \rightarrow 1 - 2x > -33 \rightarrow -2x > -33 - 1 \rightarrow$
$-2x > -34 \rightarrow x < \frac{-34}{-2} \rightarrow x < \frac{-34}{-2} \rightarrow x < 17$

42) $-3(3 - 2x) > -33 \rightarrow -9 + 6x > -33 \rightarrow 6x > -33 + 9 \rightarrow 6x > -24 \rightarrow$
$\frac{6x}{6} > \frac{-24}{6} \rightarrow x > -4$

DAY 7 — Lines and Slope

Math topics that you'll learn in this chapter:

1. Finding Slope
2. Graphing Lines Using Slope–Intercept Form
3. Writing Linear Equations
4. Finding Midpoint
5. Finding Distance of Two Points
6. Graphing Linear Inequalities

Finding Slope

★ The slope of a line represents the direction of a line on the coordinate plane.

★ A coordinate plane contains two perpendicular number lines. The horizontal line is x and the vertical line is y. The point at which the two axes intersect is called the origin. An ordered pair (x, y) shows the location of a point.

★ A line on a coordinate plane can be drawn by connecting two points.

★ To find the slope of a line, we need the equation of the line or two points on the line.

★ The slope of a line with two points A (x_1, y_1) and B (x_2, y_2) can be found by using this formula: $\frac{y_2 - y_1}{x_2 - x_1} = \frac{rise}{run}$

★ The equation of a line is typically written as $y = mx + b$ where m is the slope and b is the y-intercept.

Examples:

Example 1. Find the slope of the line through these two points:

A$(5, -5)$ and $B(9, 7)$.

Solution: Slope $= \frac{y_2 - y_1}{x_2 - x_1}$. Let (x_1, y_1) be A$(5, -5)$ and (x_2, y_2) be $B(9, 7)$.

(Remember, you can choose any point for (x_1, y_1) and (x_2, y_2)).

Then: slope $= \frac{y_2 - y_1}{x_2 - x_1} = \frac{7 - (-5)}{9 - 5} = \frac{12}{4} = 3$

The slope of the line through these two points is 3.

Example 2. Find the slope of the line with equation $y = -2x + 8$

Solution: When the equation of a line is written in the form of $y = mx + b$, the slope is m. In this line: $y = -2x + 8$, the slope is -2.

Graphing Lines Using Slope−Intercept Form

✯ Slope−intercept form of a line: given the slope m and the y−intercept (the intersection of the line and y-axis) b, then the equation of the line is:

$$y = mx + b$$

✯ To draw the graph of a linear equation in a slope-intercept form on the xy coordinate plane, find two points on the line by plugging two values for x and calculating the values of y.

✯ You can also use the slope (m) and one point to graph the line.

Example:

Sketch the graph of $y = -3x + 6$.

Solution: To graph this line, we need to find two points. When x is zero the value of y is 6. And when y is 0 the value of x is 2.

$$x = 0 \rightarrow y = -3(0) + 6 = 6$$
$$y = 0 \rightarrow 0 = -3x + 6 \rightarrow x = 2$$

Now, we have two points:
$(0, 6)$ and $(2, 0)$.
Find the points on the coordinate plane and graph the line. Remember that the slope of the line is -3.

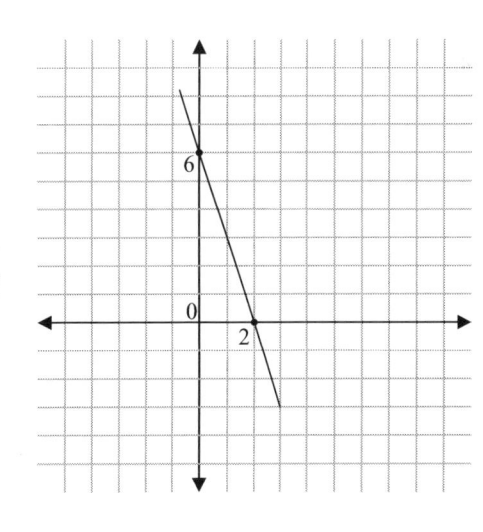

Writing Linear Equations

☆ The equation of a line in slope-intercept form: $y = mx + b$

☆ To write the equation of a line, first identify the slope.

☆ Find the y-intercept. This can be done by substituting the slope and the coordinates of a point (x, y) on the line.

Examples:

Example 1. What is the equation of the line that passes through $(-7, 2)$ and has a slope of 4?

Solution: The general slope-intercept form of the equation of a line is
$y = mx + b$, where m is the slope and b is the y-intercept.
By substitution of the given point and given slope:
$y = mx + b \to 2 = (4)(-7) + b$. So, $b = 2 + 28 = 30$, and the required equation of the line is: $y = 4x + 30$

Example 2. Write the equation of the line through two points $A(5, 2)$ and $B(3, -4)$.

Solution: First, find the slope: $Slop = \frac{y_2 - y_1}{x_2 - x_1} = \frac{-4 - 2}{3 - 5} = \frac{-6}{-2} = 3 \to m = 3$

To find the value of b, use either point and plug in the values of x and y in the equation. The answer will be the same: $y = x + b$. Let's check both points. Then:
$(5, 2) \to y = mx + b \to 2 = 3(5) + b \to b = -13$
$(3, -4) \to y = mx + b \to -4 = 3(3) + b \to b = -13$.
The y-intercept of the line is -13. The equation of the line is: $y = 3x - 13$

Example 3. What is the equation of the line that passes through $(3, -4)$ and has a slope of 2?

Solution: The general slope-intercept form of the equation of a line is
$y = mx + b$, where m is the slope and b is the y-intercept.
By substitution of the given point and given slope: $y = mx + b \to -4 = (2)(3) + b$
So, $b = -4 - 6 = -10$, and the equation of the line is: $y = 2x - 10$.

Finding Midpoint

☆ The middle of a line segment is its midpoint.

☆ The Midpoint of two endpoints A (x_1, y_1) and B (x_2, y_2) can be found using this formula: $M = \left(\frac{x_1+x_2}{2}, \frac{y_1+y_2}{2}\right)$

Examples:

Example 1. Find the midpoint of the line segment with the given endpoints. $(3, 5), (1, 3)$

Solution: Midpoint $= \left(\frac{x_1+x_2}{2}, \frac{y_1+y_2}{2}\right) \rightarrow (x_1, y_1) = (3, 5)$ and $(x_2, y_2) = (1, 3)$
Midpoint $= \left(\frac{3+1}{2}, \frac{5+3}{2}\right) \rightarrow \left(\frac{4}{2}, \frac{8}{2}\right) \rightarrow M(2, 4)$

Example 2. Find the midpoint of the line segment with the given endpoints. $(-1, 3), (9, -9)$

Solution: Midpoint $= \left(\frac{x_1+x_2}{2}, \frac{y_1+y_2}{2}\right) \rightarrow (x_1, y_1) = (-1, 3)$ and $(x_2, y_2) = (9, -9)$
Midpoint $= \left(\frac{-1+9}{2}, \frac{3+(-9)}{2}\right) \rightarrow \left(\frac{8}{2}, \frac{-6}{2}\right) \rightarrow M(4, -3)$

Example 3. Find the midpoint of the line segment with the given endpoints. $(8, 4), (-2, 6)$

Solution: Midpoint $= \left(\frac{x_1+x_2}{2}, \frac{y_1+y_2}{2}\right) \rightarrow (x_1, y_1) = (8, 4)$ and $(x_2, y_2) = (-2, 6)$
Midpoint $= \left(\frac{8-2}{2}, \frac{4+6}{2}\right) \rightarrow \left(\frac{6}{2}, \frac{10}{2}\right) \rightarrow M(3, 5)$

Example 4. Find the midpoint of the line segment with the given endpoints. $(7, -4), (-3, -8)$

Solution: Midpoint $= \left(\frac{x_1+x_2}{2}, \frac{y_1+y_2}{2}\right) \rightarrow (x_1, y_1) = (7, -4)$ and $(x_2, y_2) = (-3, -8)$
Midpoint $= \left(\frac{7-3}{2}, \frac{-4-8}{2}\right) \rightarrow \left(\frac{4}{2}, \frac{-12}{2}\right) \rightarrow M(2, -6)$

Finding Distance of Two Points

☆ Use the following formula to find the distance of two points with the coordinates A (x_1, y_1) and B (x_2, y_2):

$$d = \sqrt{(x_2 - x_1)^2 + (y_2 - y_1)^2}$$

Examples:

Example 1. Find the distance between $(5, -6)$ and $(-3, 9)$. on the coordinate plane.

Solution: Use distance of two points formula: $d = \sqrt{(x_2 - x_1)^2 + (y_2 - y_1)^2}$
$(x_1, y_1) = (5, -6)$ and $(x_2, y_2) = (-3, 9)$. Then: $d = \sqrt{(x_2 - x_1)^2 + (y_2 - y_1)^2} =$
$\sqrt{(-3 - 5)^2 + (9 - (-6))^2} = \sqrt{(-8)^2 + (15)^2} = \sqrt{64 + 225} = \sqrt{289} = 17$.
Then: $d = 17$

Example 2. Find the distance of two points $(-3, 10)$ and $(-9, 2)$

Solution: Use distance of two points formula: $d = \sqrt{(x_2 - x_1)^2 + (y_2 - y_1)^2}$
$(x_1, y_1) = (-3, 10)$ and $(x_2, y_2) = (-9, 2)$
Then: $d = \sqrt{(x_2 - x_1)^2 + (y_2 - y_1)^2} \rightarrow d = \sqrt{(-9 - (-3))^2 + (2 - 10)^2} =$
$\sqrt{(-6)^2 + (-8)^2} = \sqrt{36 + 64} = \sqrt{100} = 10$. Then: $d = 10$

Example 3. Find the distance between $(-8, 7)$ and $(4, -9)$.

Solution: Use distance of two points formula: $d = \sqrt{(x_2 - x_1)^2 + (y_2 - y_1)^2}$
$(x_1, y_1) = (-8, 7)$ and $(x_2, y_2) = (4, -9)$.
Then: $d = \sqrt{(x_2 - x_1)^2 + (y_2 - y_1)^2}$
$$d = \sqrt{(4 - (-8))^2 + (-9 - 7)^2} = \sqrt{(12)^2 + (-16)^2} = \sqrt{144 + 256} =$$
$\sqrt{400} = 20$. Then: $d = 20$

Graphing Linear Inequalities

☆ To graph a linear inequality, first draw a graph of the "equals" line.

☆ Use a dash line for less than (<) and greater than (>) signs and a solid line for less than and equal to (≤) and greater than and equal to (≥).

☆ Choose a testing point. (it can be any point on both sides of the line.)

☆ Put the value of (x, y) of that point in the inequality. If that works, that part of the line is the solution. If the values don't work, then the other part of the line is the solution.

Example:

Sketch the graph of inequality: $y > 2x - 5$

Solution: To draw the graph of $y > 2x - 5$, you first need to graph the line: $y = 2x - 5$

Since there is a greater than (>) sign, draw a dash line.

The slope is 2 and y-intercept is -5.

Then, choose a testing point and substitute the value of x and y from that point into the inequality. The easiest point to test is the origin: $(0, 0)$

$$(0, 0) \to y > 2x - 5 \to 0 > 2(0) - 5 \to 0 > -5$$

This is correct! 0 is greater than -5. So, this part of the line (on the left side) is the solution of this inequality.

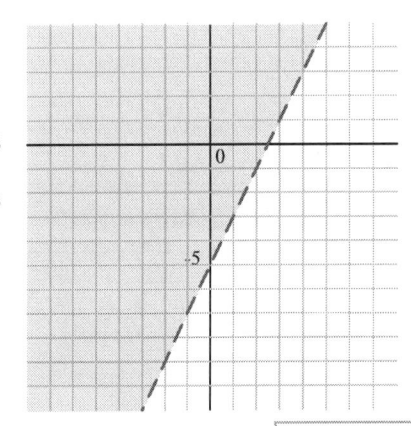

Day 7: Practices

✏ Find the slope of each line.

1) $y = x - 3$
2) $y = 3x + 4$
3) $y = -2x + 4$
4) Line through $(2, 5)$ and $(3, -4)$
5) Line through $(0, 6)$ and $(2, 4)$
6) Line through $(-2, 4)$ and $(3, -6)$

✏ Sketch the graph of each line. (Using Slope–Intercept Form)

7) $y = x + 3$

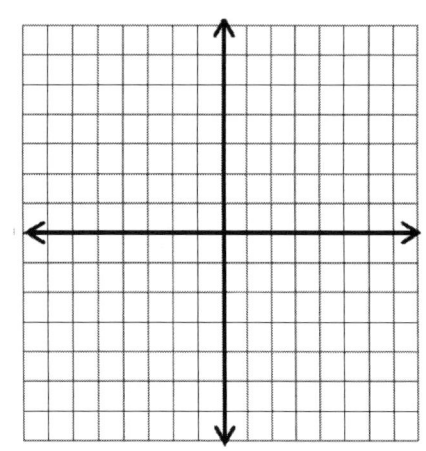

8) $y = x - 3$

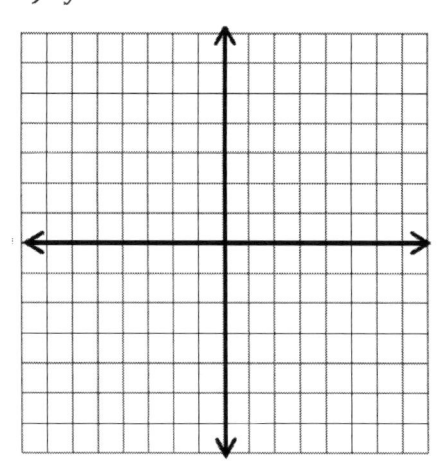

✏ Solve.

9) What is the equation of a line with slope 3 and intercept 12? _____

10) What is the equation of a line with slope 4 and passes through point $(2, 4)$? _____

11) What is the equation of a line with slope -2 and passes through point $(5, -3)$? _____

12) The slope of a line is -5 and it passes through point $(-4, 3)$. What is the equation of the line? _____

13) The slope of a line is -6 and it passes through point $(-2, -3)$. What is the equation of the line? _____

Sketch the graph of each linear inequality.

14) $y > 3x - 3$

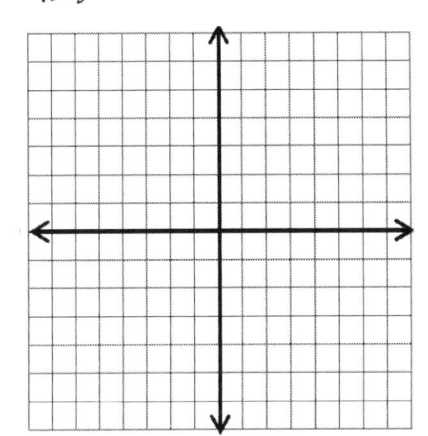

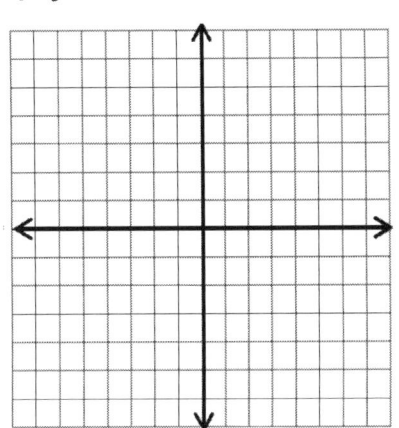

15) $y > -2x + 1$

Find the midpoint of the line segment with the given endpoints.

16) $(4, 1), (2, 3)$

17) $(3, 6), (5, 4)$

18) $(7, 1), (1, 3)$

19) $(2, 8), (2, 10)$

20) $(3, -2), (-1, 6)$

21) $(-1, -3), (1, 5)$

22) $(1, 4), (-7, 6)$

23) $(-3, 5), (7, -9)$

Find the distance between each pair of points.

24) $(-8, -1), (-4, 2)$

25) $(-15, 2), (5, -13)$

26) $(-1, 11), (-7, 3)$

27) $(0, 11), (9, 11)$

28) $(-2, 4), (3, -8)$

29) $(6, -7), (-9, 1)$

30) $(8, -4), (-4, -20)$

31) $(5, 1), (9, -2)$

32) $(-8, -17), (2, 7)$

33) $(18, 21), (-12, 5)$

Day 7: Answers

1) $y = mx + b$, the slope is m. In this line: $y = x - 3$, the slope is $m = 1$.
2) $y = 3x + 4$, the slope is $m = 3$.
3) $y = -2x + 4$, the slope is $m = -2$.
4) $(x_1, y_1) = (2, 5)$ and $(x_2, y_2) = (3, 4) \to m = \frac{y_2 - y_1}{x_2 - x_1} = \frac{4-5}{3-2} = \frac{-1}{1} = -1$
5) $(x_1, y_1) = (0, 6)$ and $(x_2, y_2) = (2, -4) \to m = \frac{y_2 - y_1}{x_2 - x_1} = \frac{-4-6}{2-0} = \frac{-10}{2} = -5$
6) $(x_1, y_1) = (-2, 4)$ and $(x_2, y_2) = (3, -6) \to m = \frac{y_2 - y_1}{x_2 - x_1} = \frac{-6-4}{3-(-2)} = \frac{-6-4}{3+2} = \frac{-10}{5} = -2$

7) $y = x + 3$
$x = 0 \to y = 0 + 3 = 3 \to (0, 3)$
$y = 0 \to 0 = x + 3 \to x = -3$
$\to (-3, 0)$
$x = 1 \to y = 1 + 3 = 4 \to (1, 4)$
$y = 1 \to 1 = x + 3 \to x = 1 - 3$
$= -2 \to (-2, 1)$

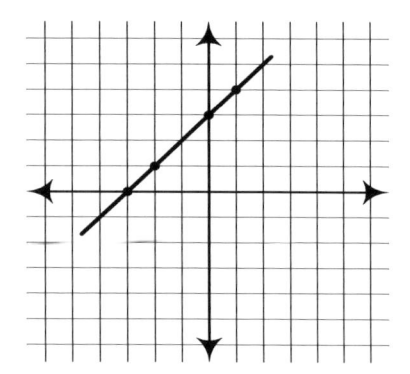

8) $y = x - 3$
$x = 0 \to y = 0 - 3 = -3 \to (0, -3)$
$y = 0 \to 0 = x - 3 \to x = 3 \to$
$\to (3, 0)$
$x = 1 \to y = 1 - 3 = -2 \to (1, -2)$
$y = 1 \to 1 = x - 3 \to x = 1 + 3 = 4$
$\to (4, 1)$

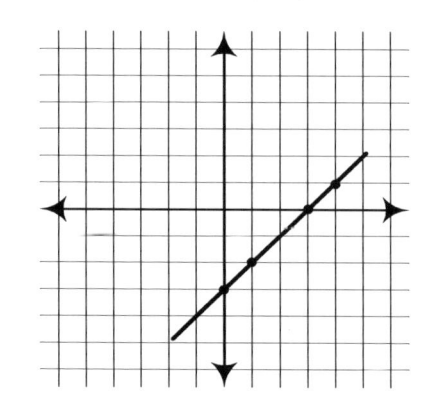

9) The general slope-intercept form of the equation of a line is $y = mx + b$, where m is the slope and b is the y-intercept $\to y = 3x + 12$
10) $y = mx + b \to 4 = 4(2) + b \to 4 = 8 + b \to b = 4 - 8 = -4 \to y = 4x - 4$
11) $y = mx + b \to -3 = -2(5) + b \to -3 = -10 + b \to b = -3 + 10 = 7 \to y = -2x + 7$
12) $y = mx + b \to 3 = -5(-4) + b \to 3 = 20 + b \to b = 3 - 20 = -17 \to y = -5x - 17$
13) $y = mx + b \to -3 = -6(-2) + b \to -3 = 12 + b \to b = -3 - 12 = -15 \to y = -6x - 15$

14) $y > 3x - 3$

$x = 0 \rightarrow y = 0 - 3 = -3 \rightarrow (0, -3)$

$y = 0 \rightarrow 0 = 3x - 3 \rightarrow 3x = 3 \rightarrow x = 1 \rightarrow (1, 0)$

The easiest point to test is the origin: $(0, 0)$

$(0,0) \rightarrow y > 3x - 3 \rightarrow 0 > 3(0) - 3 \rightarrow 0 > -3$

This is correct! 0 is greater than -3. So, this part of the line (on the left side) is the solution of this inequality.

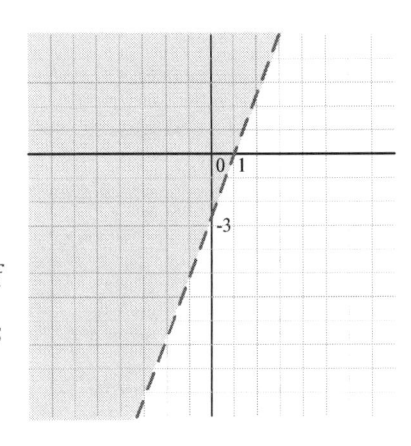

15) $y > -2x + 1$

$x = 0 \rightarrow y = 0 + 1 = 1 \rightarrow (0, 1)$

$y = 0 \rightarrow 0 = -2x + 1 \rightarrow -2x = -1 \rightarrow x = \dfrac{-1}{-2} = 0.5 \rightarrow (0.5, 0)$

The easiest point to test is the origin: $(0, 0)$

$(0,0) \rightarrow y > -2x + 1 \rightarrow 0 > -2(0) + 1 \rightarrow 0 > 1$

This is incorrect! 0 is less than 1. So, this part of the line (on the right side) is the solution of this inequality.

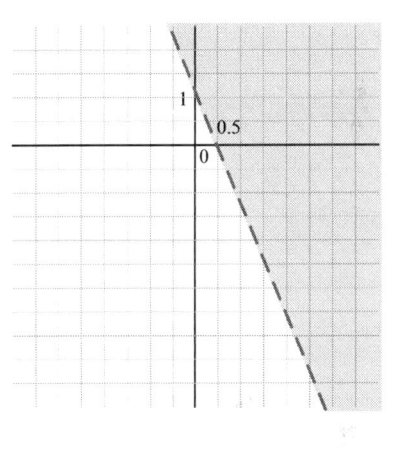

16) $M = \left(\dfrac{x_1+x_2}{2}, \dfrac{y_1+y_2}{2}\right) \rightarrow (x_1, y_1) = (4, 1)$ and $(x_2, y_2) = (2, 3) \rightarrow M = \left(\dfrac{4+2}{2}, \dfrac{1+3}{2}\right) \rightarrow \left(\dfrac{6}{2}, \dfrac{4}{2}\right) \rightarrow M(3, 2)$

17) $(x_1, y_1) = (3, 6)$ and $(x_2, y_2) = (5, 4) \rightarrow M = \left(\dfrac{3+5}{2}, \dfrac{6+4}{2}\right) \rightarrow \left(\dfrac{8}{2}, \dfrac{10}{2}\right) \rightarrow M(4, 5)$

18) $(x_1, y_1) = (7, 1)$ and $(x_2, y_2) = (1, 3) \rightarrow M = \left(\dfrac{7+1}{2}, \dfrac{1+3}{2}\right) \rightarrow \left(\dfrac{8}{2}, \dfrac{4}{2}\right) \rightarrow M(4, 2)$

19) $(x_1, y_1) = (2, 8)$ and $(x_2, y_2) = (2, 10) \rightarrow M = \left(\dfrac{2+2}{2}, \dfrac{8+10}{2}\right) \rightarrow \left(\dfrac{4}{2}, \dfrac{18}{2}\right) \rightarrow M(2, 9)$

20) $(x_1, y_1) = (3, -2)$ and $(x_2, y_2) = (-1, 6) \rightarrow M = \left(\dfrac{3-1}{2}, \dfrac{-2+6}{2}\right) \rightarrow \left(\dfrac{2}{2}, \dfrac{4}{2}\right) \rightarrow M(1, 2)$

21) $(x_1, y_1) = (-1, -3)$ and $(x_2, y_2) = (1, 5) \rightarrow M = \left(\dfrac{-1+1}{2}, \dfrac{-3+5}{2}\right) \rightarrow \left(\dfrac{0}{2}, \dfrac{2}{2}\right) \rightarrow M(0, 1)$

22) $(x_1, y_1) = (1, 4)$ and $(x_2, y_2) = (-7, 6) \rightarrow M = \left(\dfrac{1-7}{2}, \dfrac{4+6}{2}\right) \rightarrow \left(\dfrac{-6}{2}, \dfrac{10}{2}\right) \rightarrow M(-3, 5)$

23) $(x_1, y_1) = (-3, 5)$ and $(x_2, y_2) = (7, -9) \rightarrow M = \left(\dfrac{-3+7}{2}, \dfrac{5-9}{2}\right) \rightarrow \left(\dfrac{4}{2}, \dfrac{-4}{2}\right) \rightarrow M(2, -2)$

24) $(x_1, y_1) = (-8, -1)$ and $(x_2, y_2) = (-4, 2) \to d = \sqrt{(x_2 - x_1)^2 + (y_2 - y_1)^2} = \sqrt{(-4 - (-8))^2 + (2 - (-1))^2} = \sqrt{(4)^2 + (3)^2} = \sqrt{16 + 9} = \sqrt{25} = 5$

25) $(x_1, y_1) = (-15, 2)$ and $(x_2, y_2) = (5, -13) \to d = \sqrt{(x_2 - x_1)^2 + (y_2 - y_1)^2} = \sqrt{(5 - (-15))^2 + (-13 - 2)^2} = \sqrt{(20)^2 + (-15)^2} = \sqrt{400 + 225} = \sqrt{625} = 25$

26) $(x_1, y_1) = (-1, 11)$ and $(x_2, y_2) = (-7, 3) \to d = \sqrt{(x_2 - x_1)^2 + (y_2 - y_1)^2} = \sqrt{(-7 - (-1))^2 + (3 - 11)^2} = \sqrt{(-6)^2 + (-8)^2} = \sqrt{36 + 64} = \sqrt{100} = 10$

27) $(x_1, y_1) = (0, 11)$ and $(x_2, y_2) = (9, 11) \to d = \sqrt{(x_2 - x_1)^2 + (y_2 - y_1)^2} = \sqrt{(9 - 0)^2 + (11 - 11)^2} = \sqrt{(9)^2 + (0)^2} = \sqrt{81} = 9$

28) $(x_1, y_1) = (-2, 4)$ and $(x_2, y_2) = (3, -8) \to d = \sqrt{(x_2 - x_1)^2 + (y_2 - y_1)^2} = \sqrt{(3 - (-2))^2 + (-8 - 4)^2} = \sqrt{(5)^2 + (-12)^2} = \sqrt{25 + 144} = \sqrt{169} = 13$

29) $(x_1, y_1) = (6, -7)$ and $(x_2, y_2) = (-9, 1) \to d = \sqrt{(x_2 - x_1)^2 + (y_2 - y_1)^2} = \sqrt{(-9 - 6)^2 + (1 - (-7))^2} = \sqrt{(-15)^2 + (8)^2} = \sqrt{225 + 64} = \sqrt{289} = 17$

30) $(x_1, y_1) = (8, -4)$ and $(x_2, y_2) = (-4, -20) \to d = \sqrt{(x_2 - x_1)^2 + (y_2 - y_1)^2} = \sqrt{(-4 - 8)^2 + (-20 - (-4))^2} = \sqrt{(-12)^2 + (-16)^2} = \sqrt{144 + 256} = \sqrt{400} = 20$

31) $(x_1, y_1) = (5, 1)$ and $(x_2, y_2) = (9, -2) \to d = \sqrt{(x_2 - x_1)^2 + (y_2 - y_1)^2} = \sqrt{(9 - 5)^2 + (-2 - 1)^2} = \sqrt{(4)^2 + (-3)^2} = \sqrt{16 + 9} = \sqrt{25} = 5$

32) $(x_1, y_1) = (-8, -17)$ and $(x_2, y_2) = (2, 7) \to d = \sqrt{(x_2 - x_1)^2 + (y_2 - y_1)^2} = \sqrt{(2 - (-8))^2 + (7 - (-17))^2} = \sqrt{(10)^2 + (-24)^2} = \sqrt{100 + 576} = \sqrt{676} = 26$

33) $(x_1, y_1) = (18, 21)$ and $(x_2, y_2) = (-12, 5) \to d = \sqrt{(x_2 - x_1)^2 + (y_2 - y_1)^2} = \sqrt{(-12 - 18)^2 + (5 - 21)^2} = \sqrt{(-30)^2 + (-16)^2} = \sqrt{900 + 256} = \sqrt{1,156} = 34$

Effortless Math Education

DAY 8 Polynomials

Math topics that you'll learn in this chapter:

1. Simplifying Polynomials
2. Adding and Subtracting Polynomials
3. Multiplying Monomials
4. Multiplying and Dividing Monomials
5. Multiplying a Polynomial and a Monomial
6. Multiplying Binomials
7. Factoring Trinomials

Simplifying Polynomials

☆ To simplify Polynomials, find "like" terms. (they have same variables with same power).

☆ Use "FOIL". (First–Out–In–Last) for binomials:

$$(x + a)(x + b) = x^2 + (b + a)x + ab$$

☆ Add or Subtract "like" terms using order of operation.

Examples:

Example 1. Simplify this expression. $2x(3x - 4) - 6x =$

Solution: Use Distributive Property: $2x(3x - 4) = 6x^2 - 8x$

Now, combine like terms: $2x(3x - 4) - 6x = 6x^2 - 8x - 6x = 6x^2 - 14x$

Example 2. Simplify this expression. $(x + 5)(x + 7) =$

Solution: First, apply the FOIL method: $(a + b)(c + d) = ac + ad + bc + bd$

$(x + 5)(x + 7) = x^2 + 5x + 7x + 35$

Now combine like terms: $x^2 + 5x + 7x + 35 = x^2 + 12x + 35$

Example 3. Simplify this expression. $3x(-4x + 5) + 2x^2 - 5x =$

Solution: Use Distributive Property: $3x(-4x + 5) = -12x^2 + 15x$

Then: $3x(-4x + 5) + 2x^2 - 5x = -12x^2 + 15x + 2x^2 - 5x$

Now combine like terms: $-12x^2 + 2x^2 = -10x^2$, and $15x - 5x = 10x$

The simplified form of the expression: $-12x^2 + 15x + 2x^2 - 5x = -10x^2 + 10x$

Adding and Subtracting Polynomials

☆ Adding polynomials is just a matter of combining like terms, with some order of operations considerations thrown in.

☆ Be careful with the minus signs, and don't confuse addition and multiplication!

☆ For subtracting polynomials, sometimes you need to use the Distributive Property: $a(b + c) = ab + ac$, $a(b - c) = ab - ac$

Examples:

Example 1. Simplify the expressions. $(-5x^2 + 2x^3) - (-4x^3 + 2x^2) =$

Solution: First, use Distributive Property:
$-(-4x^3 + 2x^2) = 4x^3 - 2x^2$
$\rightarrow (-5x^2 + 2x^3) - (-4x^3 + 2x^2) = -5x^2 + 2x^3 + 4x^3 - 2x^2$
Now combine like terms: $2x^3 + 4x^3 = 6x^3$ and $-5x^2 - 2x^2 = -7x^2$
Then: $(x^2 - 2x^3) - (x^3 - 3x^2) = 6x^3 - 7x^2$

Example 2. Add expressions. $(2x^3 + 8) + (-x^3 + 4x^2) =$

Solution: Remove parentheses:
$$(2x^3 + 8) + (-x^3 + 4x^2) = 2x^3 + 8 - x^3 + 4x^2$$
Now combine like terms: $2x^3 + 8 - x^3 + 4x^2 = x^3 + 4x^2 + 8$

Example 3. Simplify the expressions. $(x^2 + 7x^3) - (11x^2 - 4x^3) =$

Solution: First, use Distributive Property: $-(11x^2 - 4x^3) = -11x^2 + 4x^3 \rightarrow$
$$(x^2 + 7x^3) - (11x^2 - 4x^3) = x^2 + 7x^3 - 11x^2 + 4x^3$$
Now combine like terms and write in standard form:
$x^2 + 7x^3 - 11x^2 + 4x^3 = 11x^3 - 10x^2$

Multiplying Monomials

☆ A monomial is a polynomial with just one term: Examples: $2x$ or $7y^2$.

☆ When you multiply monomials, first multiply the coefficients (a number placed before and multiplying the variable) and then multiply the variables using multiplication property of exponents.

$$x^a \times x^b = x^{a+b}$$

Examples:

Example 1. Multiply expressions. $3x^4y^5 \times 6x^2y^3$

Solution: Find the same variables and use multiplication property of exponents: $x^a \times x^b = x^{a+b}$
$x^4 \times x^2 = x^{4+2} = x^6$ and $y^5 \times y^3 = y^{5+3} = y^8$
Then, multiply coefficients and variables: $3x^4y^5 \times 6x^2y^3 = 18x^6y^8$

Example 2. Multiply expressions. $5a^5b^9 \times 3a^2b^8 =$

Solution: Use the multiplication property of exponents: $x^a \times x^b = x^{a+b}$
$a^5 \times a^2 = a^{5+2} = a^7$ and $b^9 \times b^8 = b^{9+8} = b^{17}$
Then: $5a^5b^9 \times 3a^2b^8 = 15a^7b^{17}$

Example 3. Multiply. $6x^3y^2z^4 \times 2x^2y^8z^6$

Solution: Use the multiplication property of exponents: $x^a \times x^b = x^{a+b}$
$x^3 \times x^2 = x^{3+2} = x^5$, $y^2 \times y^8 = y^{2+8} = y^{10}$ and $z^4 \times z^6 = z^{4+6} = z^{10}$
Then: $6x^3y^2z^4 \times 2x^2y^8z^6 = 12x^5y^{10}z^{10}$

Example 4. Simplify. $(2a^3b^6)(-5a^7b^{12}) =$

Solution: Use the multiplication property of exponents: $x^a \times x^b = x^{a+b}$
$a^3 \times a^7 = a^{3+7} = a^{10}$ and $b^6 \times b^{12} = b^{6+12} = b^{18}$
Then: $(2a^3b^6)(-5a^7b^{12}) = -10a^{10}b^{18}$

Multiplying and Dividing Monomials

✰ When you divide or multiply two monomials, you need to divide or multiply their coefficients and then divide or multiply their variables.

✰ In case of exponents with the same base, for Division, subtract their powers, for Multiplication, add their powers.

✰ Exponent's Multiplication and Division rules:

$$x^a \times x^b = x^{a+b}, \qquad \frac{x^a}{x^b} = x^{a-b}$$

Examples:

Example 1. Multiply expressions. $(5x^4)(3x^9) =$

Solution: Use multiplication property of exponents:
$x^a \times x^b = x^{a+b} \rightarrow x^4 \times x^9 = x^{13}$
Then: $(5x^4)(3x^9) = 15x^{13}$

Example 2. Divide expressions. $\frac{18x^3y^6}{9x^2y^4} =$

Solution: Use division property of exponents:
$\frac{x^a}{x^b} = x^{a-b} \rightarrow \frac{x^3}{x^2} = x^{3-2} = x^1$ and $\frac{y^6}{y^4} = y^{6-4} = y^2$
Then: $\frac{18x^3y^6}{9x^2y^4} = 2xy^2$

Example 3. Divide expressions. $\frac{51a^4b^{11}}{3a^2b^5}$

Solution: Use division property of exponents:
$\frac{x^a}{x^b} = x^{a-b} \rightarrow \frac{a^4}{a^2} = a^{4-2} = a^2$ and $\frac{b^{11}}{b^5} = b^{11-5} = b^6$
Then. $\frac{51a^4b^{11}}{3a^2b^5} = 17a^2b^6$

Multiplying a Polynomial and a Monomial

✯ When multiplying monomials, use the product rule for exponents.

$$x^a \times x^b = x^{a+b}$$

✯ When multiplying a monomial by a polynomial, use the distributive property.

$$a \times (b + c) = a \times b + a \times c = ab + ac$$
$$a \times (b - c) = a \times b - a \times c = ab - ac$$

Examples:

Example 1. Multiply expressions. $5x(4x + 7)$

Solution: Use Distributive Property:

$5x(4x + 7) = (5x \times 4x) + (5x \times 7) = 20x^2 + 35x$

Example 2. Multiply expressions. $y(2x^2 + 3y^2)$

Solution: Use Distributive Property:

$y(2x^2 + 3y^2) = y \times 2x^2 + y \times 3y^2 = 2x^2y + 3y^3$

Example 3. Multiply. $-2x(-x^2 + 3x + 6)$

Solution: Use Distributive Property:

$-2x(-x^2 + 3x + 6) = (-2x)(-x^2) + (-2x) \times (3x) + (-2x) \times (6) =$

Now simplify:

$(-2x)(-x^2) + (-2x) \times (3x) + (-2x) \times (6) = 2x^3 - 6x^2 - 12x$

EffortlessMath.com

Multiplying Binomials

☆ A binomial is a polynomial that is the sum or the difference of two terms, each of which is a monomial.

☆ To multiply two binomials, use the "FOIL" method. (First–Out–In–Last)

$$(x + a)(x + b) = x \times x + x \times b + a \times x + a \times b = x^2 + bx + ax + ab$$

Examples:

Example 1. Multiply Binomials. $(x - 5)(x + 4) =$

Solution: Use "FOIL". (First–Out–In–Last):

$(x - 5)(x + 4) = x^2 - 5x + 4x - 20$

Then combine like terms: $x^2 - 5x + 4x - 20 = x^2 - x - 20$

Example 2. Multiply. $(x + 3)(x + 6) =$

Solution: Use "FOIL". (First–Out–In–Last):

$(x + 3)(x + 6) = x^2 + 3x + 6x + 18$

Then simplify: $x^2 + 3x + 6x + 18 = x^2 + 9x + 18$

Example 3. Multiply. $(x - 8)(x + 4) =$

Solution: Use "FOIL". (First–Out–In–Last):

$(x - 8)(x + 4) = x^2 - 8x + 4x - 32$

Then simplify: $x^2 - 8x + 4x - 32 = x^2 - 4x - 32$

Example 4. Multiply Binomials. $(x - 6)(x - 3) =$

Solution: Use "FOIL". (First–Out–In–Last):

$(x - 6)(x - 3) = x^2 - 6x - 3x + 18$

Then combine like terms: $x^2 - 6x - 3x + 18 = x^2 - 9x + 18$

Factoring Trinomials

To factor trinomials, you can use following methods:

☆ "FOIL": $(x + a)(x + b) = x^2 + (b + a)x + ab$

☆ "Difference of Squares":
$$a^2 - b^2 = (a + b)(a - b)$$
$$a^2 + 2ab + b^2 = (a + b)(a + b)$$
$$a^2 - 2ab + b^2 = (a - b)(a - b)$$

☆ "Reverse FOIL": $x^2 + (b + a)x + ab = (x + a)(x + b)$

Examples:

Example 1. Factor this trinomial. $x^2 - 3x - 18$

Solution: Break the expression into groups. You need to find two numbers that their product is -18 and their sum is -3. (remember "Reverse FOIL": $x^2 + (b + a)x + ab = (x + a)(x + b)$). Those two numbers are 3 and -6. Then:
$$x^2 - 3x - 18 = (x^2 + 3x) + (-6x - 18)$$
Now factor out x from $x^2 + 3x$: $x(x + 3)$, and factor out -6 from $-6x - 18$: $-6(x + 3)$; Then: $(x^2 + 3x) + (-6x - 18) = x(x + 3) - 6(x + 3)$
Now factor out like term: $(x + 3)$. Then: $(x + 3)(x - 6)$

Example 2. Factor this trinomial. $2x^2 - 4x - 48$

Solution: Break the expression into groups: $(2x^2 + 8x) + (-12x - 48)$
Now factor out $2x$ from $2x^2 + 8x$: $2x(x + 4)$, and factor out -12 from $-12-48$: $-12(x + 4)$; Then: $2x(x + 4) - 12(x + 4)$, now factor out like term: $(x + 4) \to 2x(x + 4) - 12(x + 4) = (x + 4)(2x - 12)$

Day 8: Practices

✍ Simplify each polynomial.

1) $4(3x + 2) =$

2) $7(6x - 3) =$

3) $x(4x + 5) + 6x =$

4) $2x(x - 4) + 8x =$

5) $3(5x + 3) - 9x =$

6) $x(6x - 5) - 4x^2 + 11 =$

7) $-x^2 + 7 + 3x(x + 2) =$

8) $6x^2 - 7 + 3x(5x - 7) =$

✍ Add or subtract polynomials.

9) $(x^2 + 5) + (3x^2 - 2) =$

10) $(4x^2 - 5x) - (x^2 + 7x) =$

11) $(6x^3 - 2x^2) + (3x^3 - 7x^2) =$

12) $(6x^3 - 7x) - (9x^3 - 3x) =$

13) $(9x^3 + 5x^2) + (12x^2 - 7) =$

14) $(5x^3 - 8) - (2x^3 - 6x^2) =$

15) $(10x^3 + 4x) - (7x^3 - 5x) =$

16) $(12x^3 - 7x) - (3x^3 + 9x) =$

✍ Find the products. (Multiplying Monomials)

17) $5x^3 \times 6x^5 =$

18) $3x^4 \times 4x^3 =$

19) $-7a^3 b \times 3a^2 b^5 =$

20) $-5x^2 y^3 z \times 6x^6 y^4 z^5 =$

21) $-2a^3 bc \times (-4a^8 b^7) =$

22) $9u^6 t^5 \times (-2u^2 t) =$

23) $14x^2 z \times 2x^6 y^8 z =$

24) $-12x^7 y^6 z \times 3xy^8 =$

25) $-9a^2 b^3 c \times 3a^7 b^6 =$

26) $-11x^9 y^7 \times (-6x^4 y^3) =$

✍ **Simplify each expression. (Multiplying and Dividing Monomials)**

27) $(4x^3y^4)(2x^4y^3) =$

28) $(7x^2y^5)(3x^3y^6) =$

29) $(5x^9y^6)(8x^6y^9) =$

30) $(13a^4b^7)(2a^6b^9) =$

31) $\frac{54x^6y^3}{9x^4y} =$

32) $\frac{28x^5y^7}{4x^3y^4} =$

33) $\frac{32x^{17}y^{12}}{8x^{13}y^9} =$

34) $\frac{40x^7y^{19}}{5x^2y^{14}} =$

✍ **Find each product. (Multiplying a Polynomial and a Monomial)**

35) $6(4x - 2y) =$

36) $4x(5x + y) =$

37) $8x(x - 2y) =$

38) $x(3x^2 + 4x - 6) =$

39) $4x(2x^2 + 7x + 4) =$

40) $8x(3x^2 - 7x - 3) =$

✍ **Find each product. (Multiplying Binomials)**

41) $(x - 4)(x + 5) =$

42) $(x - 3)(x + 3) =$

43) $(x + 8)(x + 7) =$

44) $(x - 5)(x + 9) =$

45) $(2x + 4)(x - 6) =$

46) $(2x - 11)(x + 6) =$

✍ **Factor each trinomial.**

47) $x^2 + 2x - 15 =$

48) $x^2 - x - 42 =$

49) $x^2 - 14x + 49 =$

50) $x^2 - 7x - 60 =$

51) $2x^2 + 6x - 20 =$

52) $3x^2 + 13x - 10 =$

Effortless Math Education

EffortlessMath.com

Day 8: Answers

1) $4(3x + 2) = (4 \times 3x) + (4 \times 2) = 12x + 8$

2) $7(6x - 3) = (7 \times 6x) - (7 \times 3) = 42x - 21$

3) $x(4x + 5) + 6x = (x \times 4x) + (x \times 5) + 6x = 4x^2 + 5x + 6x = 4x^2 + 11x$

4) $2x(x - 4) + 8x = (2x \times x) + (2x \times (-4)) + 8x = 2x^2 - 8x + 8x = 2x^2$

5) $3(5x + 3) - 9x = (3 \times 5x) + (3 \times 3) - 9x = 15x - 9x + 9 = 6x + 9$

6) $x(6x - 5) - 4x^2 + 11 = (x \times 6x) + (x \times (-5)) - 4x^2 + 11 =$
   $6x^2 - 4x^2 - 5x + 11 = 2x^2 - 5x + 11$

7) $-x^2 + 7 + 3x(x + 2) = -x^2 + 7 + (3x \times x) + (3x \times 2) = -x^2 + 7 + 3x^2 + 6x =$
   $2x^2 + 6x + 7$

8) $6x^2 - 7 + 3x(5x - 7) = 6x^2 - 7 + (3x \times 5x) + (3x \times (-7)) =$
   $6x^2 - 7 + 15x^2 - 21x = 21x^2 - 21x - 7$

9) $(x^2 + 5) + (3x^2 - 2) = x^2 + 3x^2 + 5 - 2 = 4x^2 + 3$

10) $(4x^2 - 5x) - (x^2 + 7x) = 4x^2 - x^2 - 5x - 7x = 3x^2 - 12x$

11) $(6x^3 - 2x^2) + (3x^3 - 7x^2) = 6x^3 + 3x^3 - 2x^2 - 7x^2 = 9x^3 - 9x^2$

12) $(6x^3 - 7x) - (9x^3 - 3x) = 6x^3 - 9x^3 - 7x + 3x = -3x^3 - 4x$

13) $(9x^3 + 5x^2) + (12x^2 - 7) = 9x^3 + 5x^2 + 12x^2 - 7 = 9x^3 + 17x^2 - 7$

14) $(5x^3 - 8) - (2x^3 - 6x^2) = 5x^3 - 2x^3 + 6x^2 - 8 = 3x^3 + 6x^2 - 8$

15) $(10x^3 + 4x) - (7x^3 - 5x) = 10x^3 - 7x^3 + 4x + 5x = 3x^3 + 9x$

16) $(12x^3 - 7x) - (3x^3 + 9x) = 12x^3 - 3x^3 - 7x - 9x = 9x^3 - 16x$

17) $5x^3 \times 6x^5 \to 5 \times 6 = 30, \ x^3 \times x^5 = x^{3+5} = x^8 \to 5x^3 \times 6x^5 = 30x^8$

18) $3x^4 \times 4x^3 \to 3 \times 4 = 12, \ x^4 \times x^3 = x^{4+3} = x^7 \to 3x^4 \times 4x^3 = 12x^7$

19) $-7a^3b \times 3a^2b^5 \to -7 \times 3 = -21, \ a^3 \times a^2 = a^{3+2} = a^5, \ b \times b^5 = b^{1+5} = b^6 \to$
    $-7a^3b \times 3a^2b^5 = -21a^5b^6$

20) $-5x^2y^3z \times 6x^6y^4z^5 \to -5 \times 6 = -30, \ x^2 \times x^6 = x^{2+6} = x^8, y^3 \times y^4 = y^{3+4} = y^7,$
    $z \times z^5 = z^{1+5} = z^6 \to -5x^2y^3z \times 6x^6y^4z^5 = -30x^8y^7z^6$

21) $-2a^3bc \times (-4a^8b^7) \to -2 \times (-4) = 8, a^3 \times a^8 = a^{3+8} = a^{11}, b \times b^7 = b^{1+7} = b^8 \to -2a^3bc \times (-4a^8b^7) = 8a^{11}b^8c$

22) $9u^6t^5 \times (-2u^2t) \to 9 \times (-2) = -18, u^6 \times u^2 = u^{6+2} = u^8, t^5 \times t^1 = t^{5+1} = t^6 \to 9u^6t^5 \times (-2u^2t) = -18u^8t^6$

23) $14x^2z \times 2x^6y^8z \to 14 \times 2 = 28, x^2 \times x^6 = x^{2+6} = x^8, z \times z = z^{1+1} = z^2 \to 14x^2z \times 2x^6y^8z = 28x^8y^8z^2$

24) $-12x^7y^6z \times 3xy^8 \to -12 \times 3 = -36, x^7 \times x = x^{1+7} = x^8, y^6 \times y^8 = y^{6+8} = y^{14} \to -12x^7y^6z \times 3xy^8 = -36x^8y^{14}z$

25) $-9a^2b^3c \times 3a^7b^6 \to -9 \times 3 = -27, a^2 \times a^7 = a^{2+7} = a^9, b^3 \times b^6 = b^{3+6} = b^9 \to -9a^2b^3c \times 3a^7b^6 = -27a^9b^9c$

26) $-11x^9y^7 \times (-6x^4y^3) \to -11 \times (-6) = 66, x^9 \times x^4 = x^{9+4} = x^{13}, y^7 \times y^3 = y^{7+3} = y^{10} \to -11x^9y^7 \times (-6x^4y^3) = 66x^{13}y^{10}$

27) $(4x^3y^4)(2x^4y^3) \to 4 \times 2 = 8, x^3 \times x^4 = x^{3+4} = x^7, y^4 \times y^3 = y^{4+3} = y^7 \to (4x^3y^4)(2x^4y^3) = 8x^7y^7$

28) $(7x^2y^5)(3x^3y^6) \to 7 \times 3 = 21, x^2 \times x^3 = x^{2+3} = x^5, y^5 \times y^6 = y^{5+6} = y^{11} \to (7x^2y^5)(3x^3y^6) = 21x^5y^{11}$

29) $(5x^9y^6)(8x^6y^9) \to 5 \times 8 = 40, x^9 \times x^6 = x^{9+6} = x^{15}, y^6 \times y^9 = y^{6+9} = y^{15} \to (5x^9y^6)(8x^6y^9) = 40x^{15}y^{15}$

30) $(13a^4b^7)(2a^6b^9) \to 13 \times 2 = 26, a^4 \times a^6 = a^{4+6} = a^{10}, b^7 \times b^9 = b^{7+9} = b^{16} \to (13a^4b^7)(2a^6b^9) = 26a^{10}b^{16}$

31) $\frac{54x^6y^3}{9x^4y} \to \frac{54}{9} = 6, \frac{x^6}{x^4} = x^{6-4} = x^2, \frac{y^3}{y} = y^{3-1} = y^2 \to \frac{54x^6y^3}{9x^4y} = 6x^2y^2$

32) $\frac{28x^5y^7}{4x^3y^4} \to \frac{28}{4} = 7, \frac{x^5}{x^3} = x^{5-3} = x^2, \frac{y^7}{y^4} = y^{7-4} = y^3 \to \frac{28x^5y^7}{4x^3y^4} = 7x^2y^3$

33) $\frac{32x^{17}y^{12}}{8x^{13}y^9} \to \frac{32}{8} = 4, \frac{x^{17}}{x^{13}} = x^{17-13} = x^4, \frac{y^{12}}{y^9} = y^{12-9} = y^3 \to \frac{32x^{17}y^{12}}{8x^{13}y^9} = 4x^4y^3$

34) $\frac{40x^7y^{19}}{5x^2y^{14}} = \to \frac{40}{5} = 8, \frac{x^7}{x^2} = x^{7-2} = x^5, \frac{y^{19}}{y^{14}} = y^{19-14} = y^5 \to \frac{40x^7y^{19}}{5x^2y^{14}} = 8x^5y^5$

35) $6(4x - 2y) = (6 \times 4x) - (6 \times 2y) = 24x - 12y$

36) $4x(5x + y) = (4x \times 5x) + (4x \times y) = 20x^2 + 4xy$

37) $8x(x - 2y) = (8x \times x) - (8x \times 2y) = 8x^2 - 16xy$

38) $x(3x^2 + 4x - 6) = (x \times 3x^2) + (x \times 4x) + (x \times (-6)) = 3x^3 + 4x^2 - 6x$

39) $4x(2x^2 + 7x + 4) = (4x \times 2x^2) + (4x \times 7x) + (4x \times 4) = 8x^3 + 28x^2 + 16x$

40) $8x(3x^2 - 7x - 3) = (8x \times 3x^2) + (8x \times (-7x)) + (8x \times (-3)) = 24x^3 - 56x^2 - 24x$

41) $(x - 4)(x + 5) = (x \times x) + (x \times 5) + (-4 \times x) + (-4 \times 5) = x^2 + 5x - 4x - 20 = x^2 + x - 20$

42) $(x - 3)(x + 3) = (x \times x) + (x \times 3) + (-3 \times x) + (-3 \times 3) = x^2 + 3x - 3x - 9 = x^2 - 9$

43) $(x + 8)(x + 7) = (x \times x) + (x \times 7) + (8 \times x) + (8 \times 7) = x^2 + 7x + 8x + 56 = x^2 + 15x + 56$

44) $(x - 5)(x + 9) = (x \times x) + (x \times 9) + (-5 \times x) + (-5 \times 9) = x^2 + 9x - 5x - 45 = x^2 + 4x - 45$

45) $(2x + 4)(x - 6) = (2x \times x) + (2x \times (-6)) + (4 \times x) + (4 \times (-6)) = 2x^2 - 12x + 4x - 24 = 2x^2 - 8x - 24$

46) $(2x - 11)(x + 6) = (2x \times x) + (2x \times 6) + ((-11) \times x) + ((-11) \times 6) = 2x^2 + 12x - 11x - 66 = 2x^2 + x - 66$

47) $x^2 + 2x - 15 \rightarrow$ (Use this rule: $x^2 + (b + a)x + ab = (x + a)(x + b)$). Then:
$x^2 + 2x - 15 = x^2 + (5 - 3)x + (5 \times (-3)) = (x + 5)(x - 3)$

48) $x^2 - x - 42 = x^2 + (-7 + 6)x + ((-7) \times 6) = (x - 7)(x + 6)$

49) $x^2 - 14x + 49 = x^2 + (-7 - 7)x + ((-7) \times (-7)) = (x - 7)(x - 7)$

50) $x^2 - 7x - 60x^2 = x^2 + (-12 + 5)x + ((-12) \times 5) = (x - 12)(x + 5)$

51) $2x^2 + 6x - 20 = 2x^2 + (10x - 4x) - 20 = (2x^2 + 10x) + (-4x - 20) = 2x(x + 5) - 4(x + 5) = (2x - 4)(x + 5)$

52) $3x^2 + 13x - 10 = (3x^2 + 15x) + (-2x - 10) = 3x(x + 5) - 2(x + 5) = (3x - 2)(x + 5)$

Effortless Math Education

DAY 9: Geometry and Solid Figures

Math topics that you'll learn in this chapter:

1. The Pythagorean Theorem
2. Complementary and Supplementary angles
3. Parallel lines and Transversals
4. Triangles
5. Special Right Triangles
6. Polygons
7. Circles
8. Trapezoids
9. Cubes
10. Rectangle Prisms
11. Cylinder

The Pythagorean Theorem

✩ You can use the Pythagorean Theorem to find a missing side in a right triangle.

✩ In any right triangle: $a^2 + b^2 = c^2$

Examples:

Example 1. Right triangle ABC (not shown) has two legs of lengths 3 cm (AB) and 4 cm (AC). What is the length of the hypotenuse of the triangle (side BC)?

Solution: Use Pythagorean Theorem: $a^2 + b^2 = c^2$, $a = 3$ and $b = 4$

Then: $a^2 + b^2 = c^2 \rightarrow 3^2 + 4^2 = c^2 \rightarrow 9 + 16 = c^2 \rightarrow 25 = c^2 \rightarrow c = \sqrt{25} = 5$

The length of the hypotenuse is 5 cm.

Example 2. Find the hypotenuse of this triangle.

Solution: Use Pythagorean Theorem: $a^2 + b^2 = c^2$

Then: $a^2 + b^2 = c^2 \rightarrow 15^2 + 8^2 = c^2 \rightarrow 225 + 64 = c^2$

$c^2 = 289 \rightarrow c = \sqrt{289} = 17$

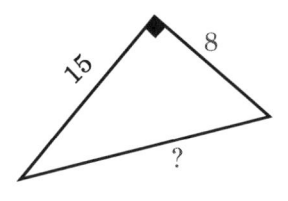

Example 3. Find the length of the missing side in this triangle.

Solution: Use Pythagorean Theorem: $a^2 + b^2 = c^2$

Then: $a^2 + b^2 = c^2 \rightarrow 20^2 + b^2 = 25^2 \rightarrow 400 + b^2 = 625 \rightarrow$

$b^2 = 625 - 400 \rightarrow b^2 = 225 \rightarrow b = \sqrt{225} = 15$

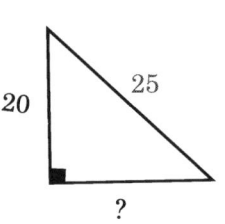

Complementary and Supplementary angles

☆ Two angles with a sum of 90 degrees are called complementary angles.

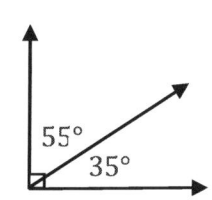

☆ Two angles with a sum of 180 degrees are Supplementary angles.

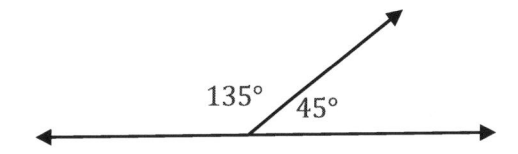

Examples:

Example 1. Find the missing angle.

Solution: Notice that the two angles form a right angle. This means that the angles are complementary, and their sum is 90. Then: $22° + x = 90° \rightarrow x = 90° - 22° = 68°$

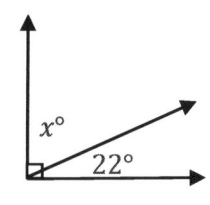

The missing angle is 68 degrees. $x = 68°$

Example 2. Angles Q and S are supplementary. What is the measure of angle Q if angle S is 45 degrees?

Solution: Q and S are supplementary $\rightarrow Q + S = 180 \rightarrow Q + 45 = 180 \rightarrow$
$$Q = 180 - 45 = 135°$$

Example 3. Angles x and y are complementary. What is the measure of angle x if angle y is 27 degrees?

Solution: Angles x and y are complementary $\rightarrow x + y = 90 \rightarrow x + 27 = 90 \rightarrow$
$$x = 90 - 27 = 63°$$

Parallel lines and Transversals

☆ When a line (transversal) intersects two parallel lines in the same plane, eight angles are formed. In the following diagram, a transversal intersects two parallel lines. Angles 1, 3, 5 and 7 are congruent. Angles 2, 4, 6, and 8 are also congruent.

☆ In the following diagram, the following angles are supplementary angles (their sum is 180):

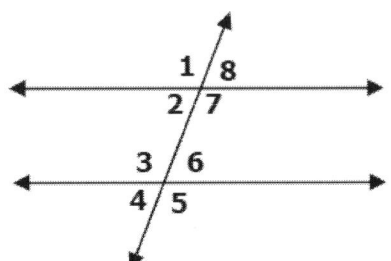

- ❖ Angles 1 and 8
- ❖ Angles 2 and 7
- ❖ Angles 3 and 6
- ❖ Angles 4 and 5

Example:

In the following diagram, two parallel lines are cut by a transversal. What is the value of x?

Solution: The two angles $2x + 4$ and $3x - 9$ are equivalent.

That is: $2x + 4 = 3x - 9$

Now, solve for x:

$2x + 4 + 9 = 3x - 9 + 9$

$\rightarrow 2x + 13 = 3x \rightarrow 2x + 13 - 2x = 3x - 2x \rightarrow$

$13 = x$

Triangles

★ In any triangle, the sum of all angles is 180 degrees.

★ Area of a triangle = $\frac{1}{2}$(base×height)

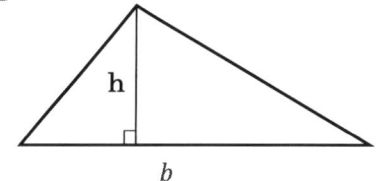

Examples:

Example 1. What is the area of this triangles?

Solution: Use the area formula:

Area= $\frac{1}{2}$(base×height)

base= 18 and height= 9, Then:

Area= $\frac{1}{2}(18 \times 9) = \frac{162}{2} = 81$

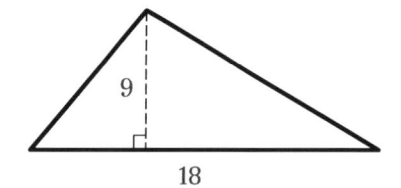

Example 2. What is the area of this triangles?

Solution: Use the area formula:

Area= $\frac{1}{2}$(base×height)

base= 18 and height= 7; Area= $\frac{1}{2}(18 \times 7) = \frac{126}{2} = 63$

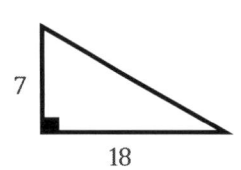

Example 3. What is the missing angle in this triangle?

Solution: In any triangle, the sum of all angles is 180 degrees. Let x be the missing angle.
Then: $63 + 84 + x = 180 \to 147 + x = 180 \to$
$$x = 180 - 147 = 33°$$
The missing angle is 33 degrees.

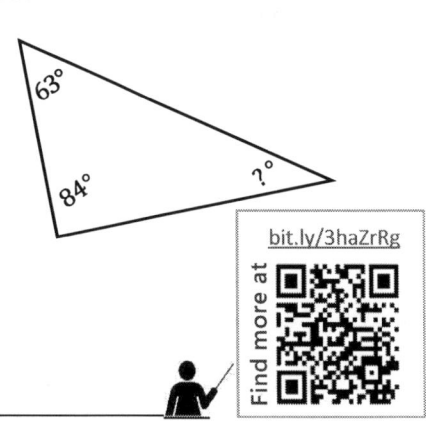

Special Right Triangles

★ A special right triangle is a triangle whose sides are in a particular ratio. Two special right triangles are $45° - 45° - 90°$ and $30° - 60° - 90°$ triangles.

★ In a special $45° - 45° - 90°$ triangle, the three angles are $45°$, $45°$ and $90°$. The lengths of the sides of this triangle are in the ratio of $1:1:\sqrt{2}$.

★ In a special triangle $30° - 60° - 90°$, the three angles are $30° - 60° - 90°$. The lengths of this triangle are in the ratio of $1:\sqrt{3}:2$.

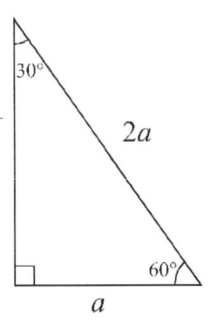

Examples:

Example 1. Find the length of the hypotenuse of a right triangle if the length of the other two sides are both 6 inches.

Solution: This is a right triangle with two equal sides. Therefore, it must be a $45° - 45° - 90°$ triangle. Two equivalent sides are 6 inches. The ratio of sides: $x:x:x\sqrt{2}$
The length of the hypotenuse is $6\sqrt{2}$ inches. $x:x:x\sqrt{2} \to 6:6:6\sqrt{2}$

Example 2. The length of the hypotenuse of a right triangle is 6 inches. What are the lengths of the other two sides if one angle of the triangle is $30°$?

Solution: The hypotenuse is 6 inches and the triangle is a $30° - 60° - 90°$ triangle. Then, one side of the triangle is 3 (it's half the side of the hypotenuse) and the other side is $3\sqrt{3}$. (it's the smallest side times $\sqrt{3}$)
$x:x\sqrt{3}:2x \to x = 3 \to x:x\sqrt{3}:2x = 3:3\sqrt{3}:6$

Polygons

★ The perimeter of a square = 4 × $side$ = 4s

★ The perimeter of a rectangle = 2($width + length$)

★ The perimeter of trapezoid = $a + b + c + d$

★ The perimeter of a regular hexagon = 6a

★ The perimeter of a parallelogram = 2($l + w$)

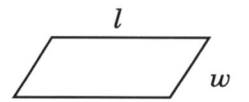

Examples:

Example 1. Find the perimeter of following regular hexagon.

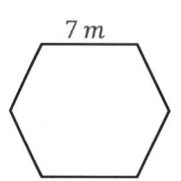

Solution: Since the hexagon is regular, all sides are equal.
Then, the perimeter of the hexagon = 6 × ($one\ side$)
The perimeter of the hexagon = 6 × ($one\ side$) = 6 × 7 = 42 m

Example 2. Find the perimeter of following trapezoid.

Solution: The perimeter of a trapezoid = $a + b + c + d$
The perimeter of the trapezoid = 9 + 7 + 13 + 7 = 36 ft

Circles

★ In a circle, variable *r* is usually used for the radius and *d* for diameter.

★ *Area of a circle* = πr^2 (π is about 3.14)

★ *Circumference of a circle* = $2\pi r$

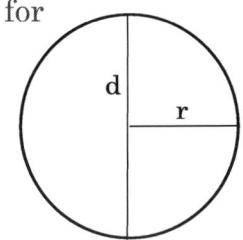

Examples:

Example 1. Find the area of this circle. ($\pi = 3.14$)

Solution:
Use area formula: $Area = \pi r^2$
$r = 5\ in \rightarrow Area = \pi(5)^2 = 25\pi$, $\pi = 3.14$
Then: $Area = 25 \times 3.14 = 78.5\ in^2$

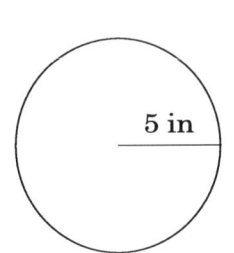

Example 2. Find the Circumference of this circle. ($\pi = 3.14$)

Solution:
Use Circumference formula: $Circumference = 2\pi r$
$r = 7\ cm \rightarrow Circumference = 2\pi(7) = 14\pi$
$\pi = 3.14$, Then: $Circumference = 14 \times 3.14 = 43.96\ cm$

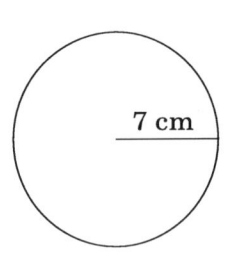

Example 3. Find the area of this circle.

Solution:
Use area formula: $Area = \pi r^2$
$r = 11\ in$, Then: $Area = \pi(11)^2 = 121\pi$, $\pi = 3.14$
$Area = 121 \times 3.14 = 379.94\ in^2$

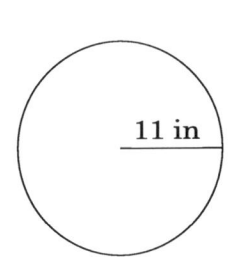

Trapezoids

★ A quadrilateral with at least one pair of parallel sides is a trapezoid.

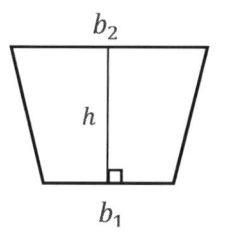

★ Area of a trapezoid = $\frac{1}{2}h(b_1 + b_2)$

Examples:

Example 1. Calculate the area of this trapezoid.

Solution:

Use area formula: $A = \frac{1}{2}h(b_1 + b_2)$

$b_1 = 5\ cm$, $b_2 = 9\ cm$ and $h = 16\ cm$

Then: $A = \frac{1}{2}(16)(9 + 5) = 8(14) = 112\ cm^2$

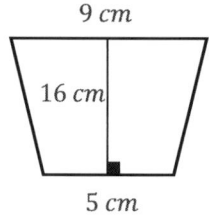

Example 2. Calculate the area of this trapezoid.

Solution:

Use area formula: $A = \frac{1}{2}h(b_1 + b_2)$

$b_1 = 8\ cm$, $b_2 = 16\ cm$ and $h = 12\ cm$

Then: $A = \frac{1}{2}(12)(8 + 16) = 144\ cm^2$

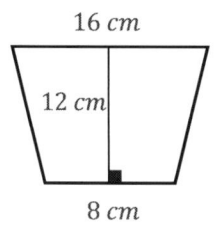

Cubes

★ A cube is a three-dimensional solid object bounded by six square sides.

★ Volume is the measure of the amount of space inside of a solid figure, like a cube, ball, cylinder or pyramid.

★ The volume of a cube = $(one\ side)^3$

★ The surface area of a cube = $6 \times (one\ side)^2$

Examples:

Example 1. Find the volume and surface area of this cube.

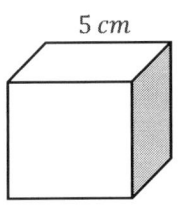

Solution: Use volume formula: $volume = (one\ side)^3$
Then: $volume = (one\ side)^3 = (5)^3 = 125\ cm^3$
Use surface area formula:
surface area of a cube: $6(one\ side)^2 = 6(5)^2 = 6(25) = 150\ cm^2$

Example 2. Find the volume and surface area of this cube.

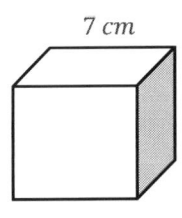

Solution: Use volume formula: $volume = (one\ side)^3$
Then: $volume = (one\ side)^3 = (7)^3 = 343\ cm^3$
Use surface area formula:
surface area of a cube: $6(one\ side)^2 = 6(7)^2 = 6(49) = 294\ cm^2$

Example 3. Find the volume and surface area of this cube.

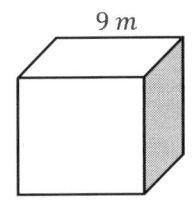

Solution: Use volume formula: $volume = (one\ side)^3$
Then: $volume = (one\ side)^3 = (9)^3 = 729\ m^3$
Use surface area formula:
$$surface\ area\ of\ a\ cube: 6(one\ side)^2 = 6(9)^2$$
$$= 6(81) = 486\ m^2$$

Rectangular Prisms

✩ A rectangular prism is a solid 3-dimensional object with six rectangular faces.

✩ The volume of a Rectangular prism = Length × Width × Height

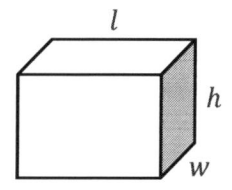

$Volume = l \times w \times h$

$Surface\ area = 2 \times (wh + lw + lh)$

Examples:

Example 1. Find the volume and surface area of this rectangular prism.

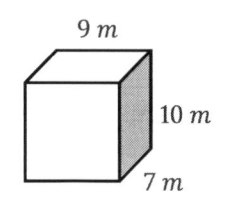

Solution: Use volume formula: $Volume = l \times w \times h$

Then: $Volume = 9 \times 7 \times 10 = 630\ m^3$

Use surface area formula: $Surface\ area = 2 \times (wh + lw + lh)$

Then: $Surface\ area = 2 \times ((7 \times 10) + (9 \times 7) + (9 \times 10))$

$= 2 \times (70 + 63 + 90) = 2 \times (223) = 446\ m^2$

Example 2. Find the volume and surface area of this rectangular prism.

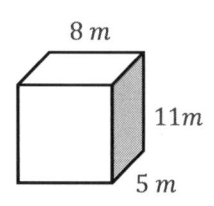

Solution: Use volume formula: $Volume = l \times w \times h$

Then: $Volume = 8 \times 5 \times 11 = 440\ m^3$

Use surface area formula: $Surface\ area = 2 \times (wh + lw + lh)$

Then: $Surface\ area = 2 \times ((5 \times 11) + (8 \times 5) + (8 \times 11))$

$= 2 \times (55 + 40 + 88) = 2 \times (183) = 366\ m^2$

Cylinder

★ A cylinder is a solid geometric figure with straight parallel sides and a circular or oval cross-section.

★ Volume of a Cylinder = $\pi(radius)^2 \times height$, $\pi \approx 3.14$

★ Surface area of a cylinder = $2\pi r^2 + 2\pi rh$

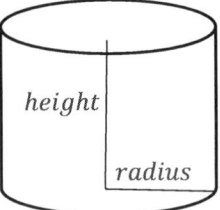

Examples:

Example 1. Find the volume and Surface area of the follow Cylinder.

Solution: Use volume formula:
Volume = $\pi(radius)^2 \times height$
Then: Volume = $\pi(6)^2 \times 12 = 36\pi \times 12 = 432\pi$
$\pi = 3.14$, then: Volume = $432\pi = 432 \times 3.14 = 1,356.48\ cm^3$
Use surface area formula: Surface area = $2\pi r^2 + 2\pi rh$
Then: $2\pi(6)^2 + 2\pi(6)(12) = 2\pi(36) + 2\pi(72) = 72\pi + 144\pi = 216\pi$
$\pi = 3.14$, Then: Surface area = $216 \times 3.14 = 678.24\ cm^2$

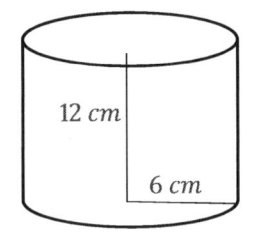

Example 2. Find the volume and Surface area of the follow Cylinder.

Solution: Use volume formula:
Volume = $\pi(radius)^2 \times height$
Then: Volume = $\pi(2)^2 \times 5 = 4\pi \times 5 = 20\pi$
$\pi = 3.14$, Then: Volume = $20\pi = 62.8\ cm^3$
Use surface area formula: Surface area = $2\pi r^2 + 2\pi rh$
Then: $= 2\pi(2)^2 + 2\pi(2)(5) = 2\pi(4) + 2\pi(10) = 8\pi + 20\pi = 28\pi$
$\pi = 3.14$, then: Surface area = $28 \times 3.14 = 87.92\ cm^2$

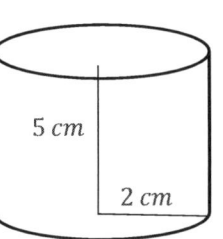

Day 9: Practices

✎ Find the missing side?

1) right triangle, legs 6 and ?, hypotenuse 10

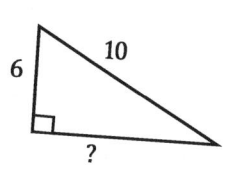

2) right triangle, legs 15 and 8, hypotenuse ?

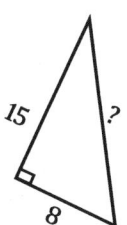

3) right triangle, legs ? and 12, hypotenuse 13

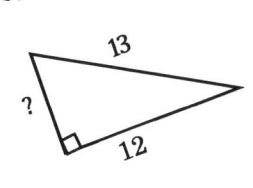

4) right triangle, legs 3 and 4, hypotenuse ?

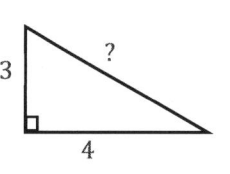

✎ Find the measure of the unknown angle in each triangle.

5) 85°, 40°, ?°

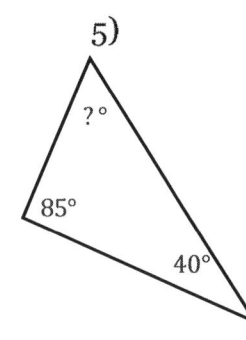

6) 58°, 42°, ?°

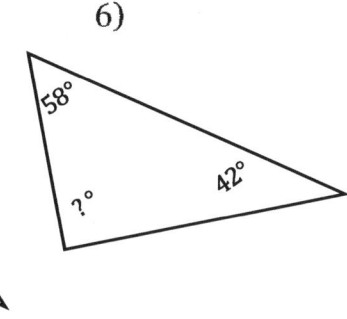

7) 71°, 18°, ?°

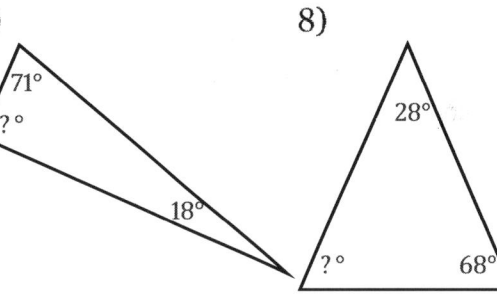

8) 28°, 68°, ?°

✎ Find the area of each triangle.

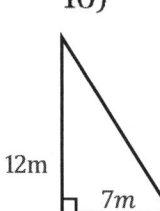

9) right triangle, legs 5 ft and 8 ft

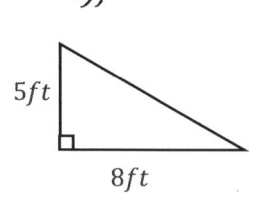

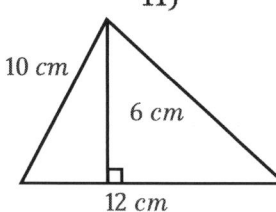

10) right triangle, legs 12 m and 7 m

11) base 12 cm, height 6 cm (slant 10 cm)

12) base 16 in, height 12 in (slant 14 in)

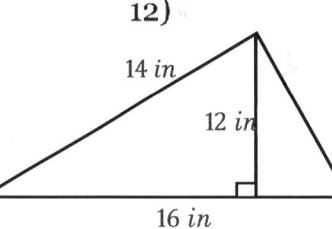

✎ Find the perimeter or circumference of each shape.

13) rhombus with sides 14 cm, 16 cm, 14 cm, 16 cm

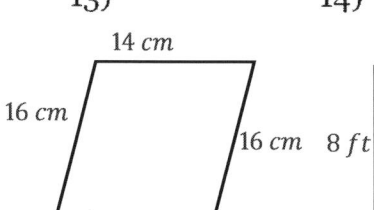

14) rectangle 10 ft by 8 ft

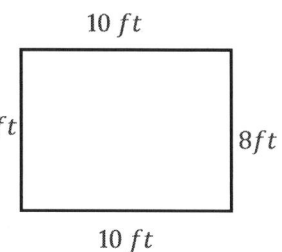

15) circle with radius 9 in

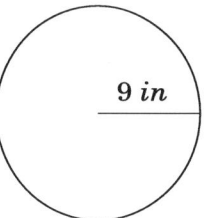

16) regular hexagon, side 7 m

Effortless Math Education

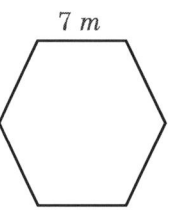

Find the area of each trapezoid.

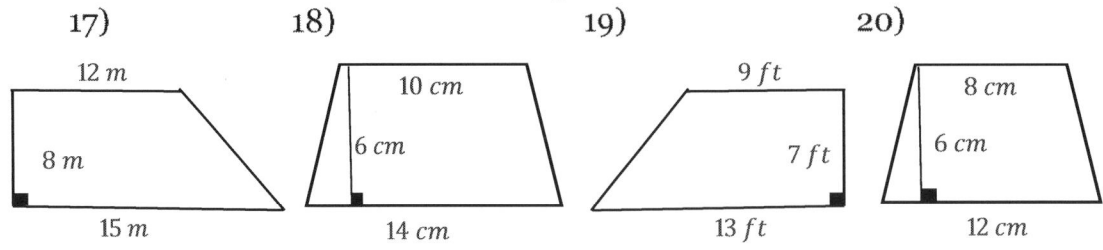

17) 12 m (top), 8 m (height), 15 m (bottom)

18) 10 cm (top), 6 cm (height), 14 cm (bottom)

19) 9 ft (top), 7 ft (height), 13 ft (bottom)

20) 8 cm (top), 6 cm (height), 12 cm (bottom)

Find the volume of each cube.

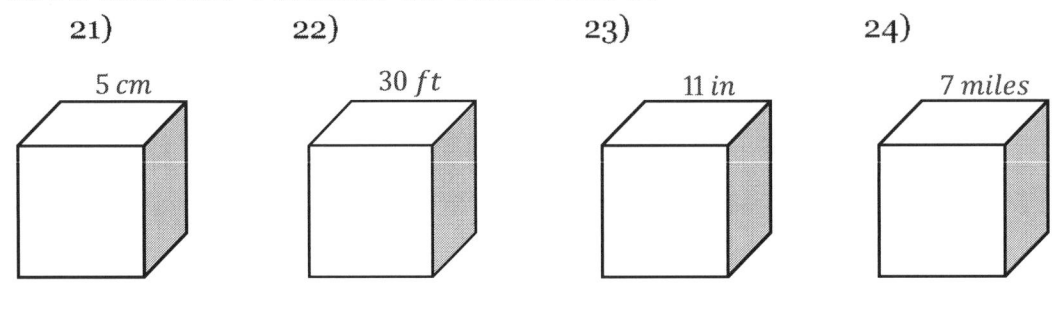

21) 5 cm

22) 30 ft

23) 11 in

24) 7 miles

Find the volume of each Rectangular Prism.

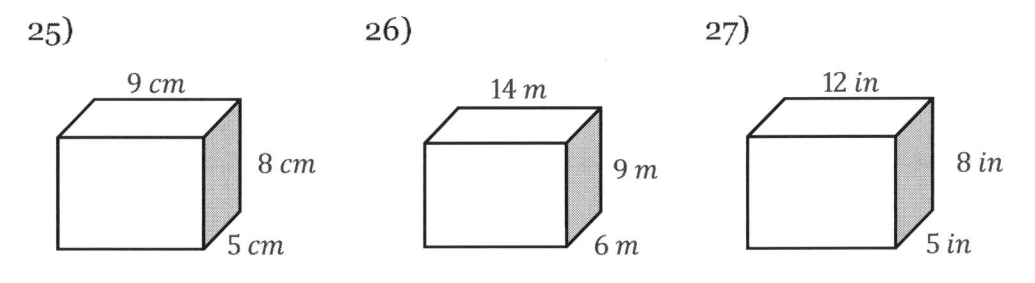

25) 9 cm, 8 cm, 5 cm

26) 14 m, 9 m, 6 m

27) 12 in, 8 in, 5 in

Find the volume of each Cylinder. Round your answer to the nearest tenth. ($\pi = 3.14$)

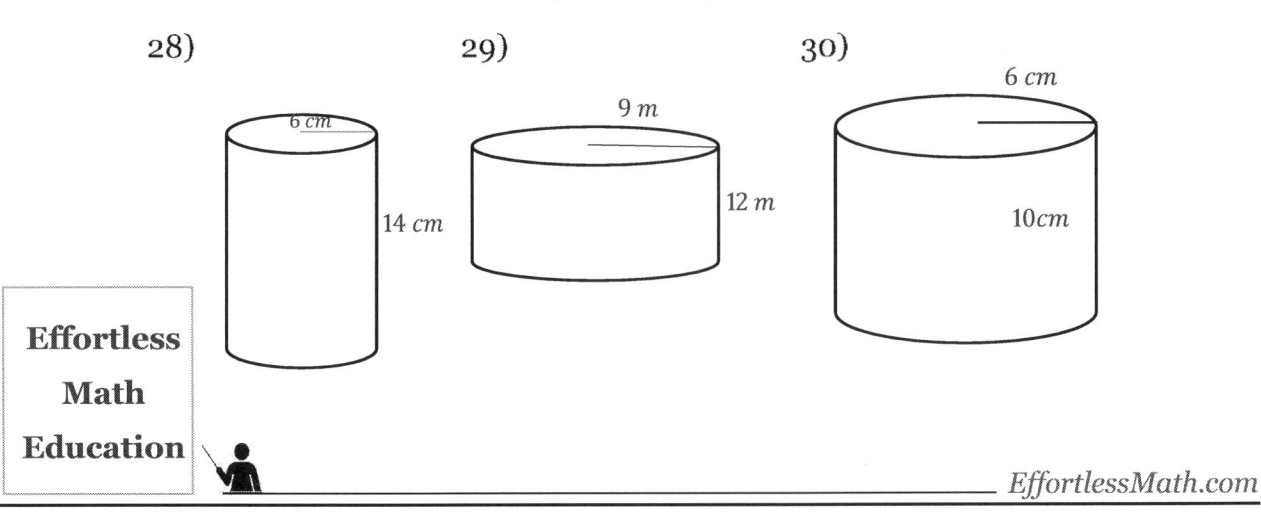

28) 6 cm (radius), 14 cm (height)

29) 9 m (radius), 12 m (height)

30) 6 cm (radius), 10 cm (height)

Day 9: Answers

1) Use Pythagorean Theorem: $a^2 + b^2 = c^2$, $a = 6$ and $c = 10$, Then: $6^2 + b^2 = 10^2 \to 36 + b^2 = 100 \to b^2 = 100 - 36 = 64 \to b = \sqrt{64} = 8$

2) Use Pythagorean Theorem: $a^2 + b^2 = c^2$, $a = 15$ and $b = 8$, Then: $15^2 + 8^2 = c^2 \to 225 + 64 = c^2 \to c^2 = 289 \to c = \sqrt{289} = 17$

3) Use Pythagorean Theorem: $a^2 + b^2 = c^2$, $a = 12$ and $c = 13$, Then: $12^2 + b^2 = 13^2 \to 144 + b^2 = 169 \to b^2 = 169 - 144 = 25 \to b = \sqrt{25} = 5$

4) Use Pythagorean Theorem: $a^2 + b^2 = c^2$, $a = 3$ and $b = 4$, Then: $3^2 + 4^2 = c^2 \to 9 + 16 = c^2 \to c^2 = 25 \to c = \sqrt{25} = 5$

5) In any triangle, the sum of all angles is 180 degrees. Then: $85 + 40 + x = 180 \to 125 + x = 180 \to x = 180 - 125 = 55$

6) $58 + 42 + x = 180 \to 100 + x = 180 \to x = 180 - 100 = 80$

7) $71 + 18 + x = 180 \to 89 + x = 180 \to x = 180 - 89 = 91$

8) $28 + 68 + x = 180 \to 96 + x = 180 \to x = 180 - 96 = 84$

9) Area $= \frac{1}{2}$(base×height), base= 8 and height = 5; Area $= \frac{1}{2}(8 \times 5) = \frac{40}{2} = 20 \ ft^2$

10) Base $= 7 \ m$ and height $= 12 \ m$; Area $= \frac{1}{2}(12 \times 7) = \frac{84}{2} = 42 \ m^2$

11) Base $= 12 \ cm$ and height $= 6 \ cm$; Area $= \frac{1}{2}(12 \times 6) = \frac{72}{2} = 36 \ cm^2$

12) Base $= 16 \ in$ and height $= 12 \ in$; Area $= \frac{1}{2}(16 \times 12) = \frac{192}{2} = 96 \ in^2$

13) The perimeter of a parallelogram $= 2(l + w) \to l = 14 \ cm, w = 16 \ cm$. Then: $2(l + w) = 2(14 + 16) = 60 \ cm$

14) The perimeter of a rectangle $= 2(w + l) \to l = 10 \ ft, w = 8 \ ft$. Then: $2(l + w) = 2(10 + 8) = 36 \ ft$

15) Circumference of a circle $= 2\pi r \to r = 9 \ in$. Then: $2\pi \times 9 = 18 \times 3.14 = 56.52 \ in$

16) The perimeter of a regular hexagon $= 6a \to a = 7 \ m$. Then: $6 \times 7 = 42 \ m$

Effortless Math Education

EffortlessMath.com

17) The area of a trapezoid $= \frac{1}{2}h(b_1 + b_2) \to b_1 = 12\,m$, $b_2 = 15\,m$ and $h = 8m$.
Then: $A = \frac{1}{2}(8)(12 + 15) = 108\,m^2$

18) $b_1 = 10\,cm$, $b_2 = 14\,cm$ and $h = 6\,cm$. Then: $A = \frac{1}{2}(6)(10 + 14) = 72\,cm^2$

19) $b_1 = 9\,ft$, $b_2 = 13\,ft$ and $h = 7\,ft$. Then: $A = \frac{1}{2}(7)(9 + 13) = 77\,ft^2$

20) $b_1 = 8\,cm$, $b_2 = 12\,cm$ and $h = 6\,cm$. Then: $A = \frac{1}{2}(6)(8 + 12) = 60\,cm^2$

21) The volume of a cube $= (one\ side)^3 = (5)^3 = 125\,cm^3$

22) The volume of a cube $= (one\ side)^3 = (30)^3 = 27{,}000\,ft^3$

23) The volume of a cube $= (one\ side)^3 = (11)^3 = 1{,}331\,in^3$

24) The volume of a cube $= (one\ side)^3 = (7)^3 = 343\,miles^3$

25) The volume of a Rectangular prism $= l \times w \times h \to l = 9\,cm, w = 5\,cm, h = 8\,cm$.
Then: $V = 9 \times 5 \times 8 = 360\,cm^3$

26) $V = l \times w \times h \to l = 14\,m, w = 6\,m, h = 9\,m$. Then: $V = 14 \times 6 \times 9 = 756\,m^3$

27) $V = l \times w \times h \to l = 12\,in, w = 5\,in, h = 8\,in$. Then: $V = 12 \times 5 \times 8 = 480\,in^3$

28) Volume of a Cylinder $= \pi(r)^2 \times h \to r = 6\,cm, h = 14\,cm$
Then: $\pi(6)^2 \times 14 = 3.14 \times 36 \times 14 = 1{,}582.56 \approx 1{,}582.6$

29) Volume of a Cylinder $= \pi(r)^2 \times h \to r = 9\,m, h = 12\,m$
Then: $\pi(9)^2 \times 12 = 3.14 \times 81 \times 12 = 3{,}052.08 \approx 3{,}052.1$

30) Volume of a Cylinder $= \pi(r)^2 \times h \to r = 6\,cm, h = 10\,cm$
Then: $\pi(6)^2 \times 10 = 3.14 \times 36 \times 10 = 1{,}130$

Day 10: Statistics and Functions

Math topics that you'll learn in this chapter:

1. Mean, Median, Mode, and Range of the Given Data
2. Pie Graph
3. Probability Problems
4. Permutations and Combinations
5. Function Notation and Evaluation
6. Adding and Subtracting Functions
7. Multiplying and Dividing Functions
8. Compositions of Functions

Mean, Median, Mode, and Range of the Given Data

☆ **Mean:** $\frac{\text{sum of the data}}{\text{total number of data entires}}$

☆ **Mode:** the value in the list that appears most often

☆ **Median:** is the middle number of a group of numbers arranged in order by size.

☆ **Range:** the difference of the largest value and smallest value in the list

Examples:

Example 1. What is the mode of these numbers? 4, 7, 8, 7, 8, 9, 8, 5

Solution: Mode: the value in the list that appears most often.
Therefore, the mode is number 8. There are three number 8 in the data.

Example 2. What is the median of these numbers? 6, 11, 15, 10, 17, 20, 7

Solution: Write the numbers in order: 6, 7, 10, 11, 15, 17, 20
The median is the number in the middle. Therefore, the median is 11.

Example 3. What is the mean of these numbers? 8, 5, 3, 7, 6, 4, 9

Solution: Mean: $\frac{\text{sum of the data}}{\text{total number of data entires}} = \frac{8+5+3+7+6+4+9}{7} = \frac{42}{7} = 6$

Example 4. What is the range in this list? 9, 2, 5, 10, 15, 22, 7

Solution: Range is the difference of the largest value and smallest value in the list. The largest value is 22 and the smallest value is 2.
Then: $22 - 2 = 20$

Pie Graph

★ A Pie Graph (Pie Chart) is a circle chart divided into sectors, each sector represents the relative size of each value.

★ Pie charts represent a snapshot of how a group is broken down into smaller pieces.

Example:

A library has 650 books that include Mathematics, Physics, Chemistry, English and History. Use the following graph to answer the questions.

Example 1. What is the number of Mathematics books?

Solution: Number of total books = 650
Percent of Mathematics books = 32%
Then, the number of Mathematics books: 32% × 650 = 0.32 × 650 = 208

Example 2. What is the number of History books?

Solution: Number of total books = 650
Percent of History books = 10%
Then: 0.10 × 650 = 65

Example 3. What is the number of English books in the library?

Solution: Number of total books = 650
Percent of English books = 14%
Then: 0.14 × 650 = 91

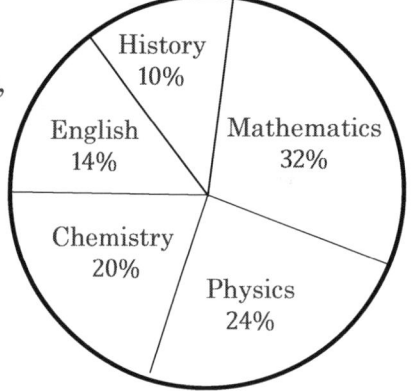

Probability Problems

★ Probability is the likelihood of something happening in the future. It is expressed as a number between zero (can never happen) to 1 (will always happen).

★ Probability can be expressed as a fraction, a decimal, or a percent.

★ Probability formula: $Probability = \frac{number\ of\ desired\ outcomes}{number\ of\ total\ outcomes}$

Examples:

Example 1. Anita's trick–or–treat bag contains 8 pieces of chocolate, 16 suckers, 22 pieces of gum and 20 pieces of licorice. If she randomly pulls a piece of candy from her bag, what is the probability of her pulling out a piece of gum?

Solution: $Probability = \frac{number\ of\ desired\ outcomes}{number\ of\ total\ outcomes}$

Probability of pulling out a piece of gum $= \frac{22}{8+16+22+20} = \frac{22}{66} = \frac{1}{3}$

Example 2. A bag contains 25 balls: five green, eight black, seven blue, a brown, a red and three white. If 24 balls are removed from the bag at random, what is the probability that a red ball has been removed?

Solution: If 24 balls are removed from the bag at random, there will be one ball in the bag. The probability of choosing a red ball is 1 out of 25. Therefore, the probability of not choosing a red ball is 24 out of 25 and the probability of having not a red ball after removing 24 balls is the same. The answer is: $\frac{24}{25}$

Permutations and Combinations

☆ **Factorials** are products, indicated by an exclamation mark. For example, $4! = 4 \times 3 \times 2 \times 1$ (Remember that 0! is defined to be equal to 1)

☆ **Permutations**: The number of ways to choose a sample of k elements from a set of n distinct objects where order does matter, and replacements are not allowed. For a permutation problem, use this formula:

$$nP_k = \frac{n!}{(n-k)!}$$

☆ **Combination**: The number of ways to choose a sample of r elements from a set of n distinct objects where order does not matter, and replacements are not allowed. For a combination problem, use this formula:

$$nC_r = \frac{n!}{r!\,(n-r)!}$$

Examples:

Example 1. How many ways can the first and second place be awarded to 6 people?

Solution: Since the order matters, (the first and second place are different!) we need to use permutation formula where n is 6 and k is 2.
Then: $\frac{n!}{(n-k)!} = \frac{6!}{(6-2)!} = \frac{6!}{4!} = \frac{6 \times 5 \times 4!}{4!}$, remove 4! from both sides of the fraction. Then: $\frac{6 \times 5 \times 4!}{4!} = 6 \times 5 = 30$

Example 2. How many ways can we pick a team of 4 people from a group of 9?

Solution: Since the order doesn't matter, we need to use a combination formula where n is 9 and r is 4.
Then: $\frac{n!}{r!\,(n-r)!} = \frac{9!}{4!\,(9-4)!} = \frac{9!}{4!\,(5)!} = \frac{9 \times 8 \times 7 \times 6 \times 5!}{4!\,(5)!} = \frac{9 \times 8 \times 7 \times 6}{4 \times 3 \times 2 \times 1} = \frac{3{,}024}{24} = 126$

Function Notation and Evaluation

☆ Functions are mathematical operations that assign unique outputs to given inputs.

☆ Function notation is the way a function is written. It is meant to be a precise way of giving information about the function without a rather lengthy written explanation.

☆ The most popular function notation is $f(x)$ which is read "f of x". Any letter can name a function. for example: $g(x)$, $h(x)$, etc.

☆ To evaluate a function, plug in the input (the given value or expression) for the function's variable (place holder, x).

Examples:

Example 1. Evaluate: $f(x) = 2x + 9$, find $f(5)$

Solution: Substitute x with 5:
Then: $f(x) = 2x + 9 \to f(5) = 2(5) + 9 \to f(5) = 19$

Example 2. Evaluate: $w(x) = 5x - 4$, find $w(2)$.

Solution: Substitute x with 2:
Then: $w(x) = 5x - 4 \to w(2) = 5(2) - 4 = 10 - 4 = 6$

Example 3. Evaluate: $f(x) = 5x^2 + 8$, find $f(-2)$.

Solution: Substitute x with -2:
Then: $f(x) = 5x^2 + 8 \to f(-2) = 5(-2)^2 + 8 \to f(-2) = 20 + 8 = 28$

Example 4. Evaluate: $h(x) = 3x^2 - 4$, find $h(3a)$.

Solution: Substitute x with $3a$:
Then: $h(x) = 3x^2 - 4 \to h(3a) = 3(3a)^2 - 4 \to h(3a) = 3(9a^2) - 4 = 27a^2 - 4$

Adding and Subtracting Functions

★ Just like we can add and subtract numbers and expressions, we can add or subtract functions and simplify or evaluate them. The result is a new function.

★ For two functions $f(x)$ and $g(x)$, we can create two new functions:

$$(f+g)(x) = f(x) + g(x) \text{ and } (f-g)(x) = f(x) - g(x)$$

Examples:

Example 1. $g(x) = 4x - 3$, $f(x) = x + 6$, Find: $(g + f)(x)$

Solution: $(g + f)(x) = g(x) + f(x)$
Then: $(g + f)(x) = (4x - 3) + (x + 6) = 4x - 3 + x + 6 = 5x + 3$

Example 2. $f(x) = 2x - 7$, $g(x) = x - 9$, Find: $(f - g)(x)$

Solution: $(f - g)(x) = f(x) - g(x)$
Then: $(f - g)(x) = (2x - 7) - (x - 9) = 2x - 7 - x + 9 = x + 2$

Example 3. $g(x) = 2x^2 + 6$, $f(x) = x - 3$, Find: $(g + f)(x)$

Solution: $(g + f)(x) = g(x) + f(x)$
Then: $(g + f)(x) = (2x^2 + 6) + (x - 3) = 2x^2 + x + 3$

Example 4. $f(x) = 2x^2 - 1$, $g(x) = 4x + 3$, Find: $(f - g)(2)$

Solution: $(f - g)(x) = f(x) - g(x)$
Then: $(f - g)(x) = (2x^2 - 1) - (4x + 3) = 2x^2 - 1 - 4x - 3 = 2x^2 - 4x - 4$
Substitute x with 2: $(f - g)(2) = 2(2)^2 - 4(2) - 4 = 8 - 8 - 4 = -4$

Multiplying and Dividing Functions

☆ Just like we can multiply and divide numbers and expressions, we can multiply and divide two functions and simplify or evaluate them.

☆ For two functions $f(x)$ and $g(x)$, we can create two new functions:
$$(f.g)(x) = f(x).g(x) \text{ and } \left(\frac{f}{g}\right)(x) = \frac{f(x)}{g(x)}$$

Examples:

Example 1. $g(x) = x + 2$, $f(x) = x + 5$, Find: $(g.f)(x)$

Solution:
$$(g.f)(x) = g(x).f(x) = (x+2)(x+5) = x^2 + 5x + 2x + 10 = x^2 + 7x + 10$$

Example 2. $f(x) = x + 4$, $h(x) = x - 16$, Find: $\left(\frac{f}{h}\right)(x)$

Solution: $\left(\frac{f}{h}\right)(x) = \frac{f(x)}{h(x)} = \frac{x+4}{x-16}$

Example 3. $g(x) = x + 5$, $f(x) = x - 2$, Find: $(g.f)(3)$

Solution: $(g.f)(x) = g(x).f(x) = (x+5)(x-2) = x^2 - 2x + 5x - 10$
$$g(x).f(x) = x^2 + 3x - 10$$
Substitute x with 3: $(g.f)(3) = (3)^2 + 3(3) - 10 = 9 + 9 - 10 = 8$

Example 4. $f(x) = 2x + 2$, $h(x) = x - 3$, Find: $\left(\frac{f}{h}\right)(4)$

Solution: $\left(\frac{f}{h}\right)(x) = \frac{f(x)}{h(x)} = \frac{2x+2}{x-3}$
Substitute x with 4: $\left(\frac{f}{h}\right)(4) = \frac{2x+2}{x-3} = \frac{2(4)+2}{4-3} = \frac{10}{1} = 10$

Composition of Functions

★ "Composition of functions" simply means combining two or more functions in a way where the output from one function becomes the input for the next function.

★ The notation used for composition is: $(fog)(x) = f(g(x))$ and is read "f composed with g of x" or "f of g of x".

Examples:

Example 1. Using $f(x) = x + 5$ and $g(x) = 7x$, find: $(fog)(x)$

Solution: $(fog)(x) = f(g(x))$. Then: $(fog)(x) = f(g(x)) = f(7x)$

Now find $f(7x)$ by substituting x with $7x$ in $f(x)$ function.

Then: $f(x) = x + 5; (x \to 7x) \to f(7x) = 7x + 5$

Example 2. Using $f(x) = 2x - 3$ and $g(x) = x - 5$, find: $(gof)(3)$

Solution: $(fog)(x) = f(g(x))$. Then: $(gof)(x) = g(f(x)) = g(2x - 3)$,

Now substitute x in $g(x)$ by $(2x - 3)$.

Then: $g(2x - 3) = (2x - 3) - 5 = 2x - 8$

Substitute x with 3: $(gof)(3) = g(f(x)) = 2x - 8 = 2(3) - 8 = -2$

Example 3. Using $f(x) = 3x^2 - 7$ and $g(x) = 2x + 1$, find: $f(g(2))$

Solution: First, find $g(2)$: $g(x) = 2x + 1 \to g(2) = 2(2) + 1 = 5$

Then: $f(g(2)) = f(5)$. Now, find $f(5)$ by substituting x with 5 in $f(x)$ function.

$f(g(2)) = f(5) = 3(5)^2 - 7 = 3(25) - 7 = 68$

Day 10: Practices

✎ **Find the values of the Given Data.**

1) 9, 10, 9, 8, 11

   Mode: _____   Range: _____

2) 6, 8, 1, 4, 7, 6, 10

   Mean: _____   Median: _____

✎ **The circle graph below shows all Bob's expenses for last month. Bob spent $675 on his Rent last month.**

Bob's last month expenses
- Rent 45%
- Foods 10%
- Bills 28%
- Car 14%
- Others 3%

3) How much did Bob's total expenses last month? _____

4) How much did Bob spend for his bills last month? _____

5) How much did Bob spend for his car last month? _____

✎ **Solve.**

6) Bag A contains 9 red marbles and 6 green marbles. Bag B contains 6 black marbles and 9 orange marbles. What is the probability of selecting a red marble at random from bag A? What is the probability of selecting a black marble at random from Bag B?

   _____   _____

7) Jason is planning for his vacation. He wants to go to museum, go to the beach, and play volleyball. How many different ways of ordering are there for him? _____

8) In how many ways can a team of 8 basketball players choose a captain and co-captain? _____

9) How many ways can you give 9 balls to your 12 friends? _____

Evaluate each function.

10) $g(n) = 3n + 7$, find $g(3)$

11) $h(x) = 4n - 7$, find $h(5)$

12) $y(n) = 12 - 3n$, find $y(8)$

13) $b(n) = -10 - 5n$, find $b(8)$

14) $g(x) = -7x + 6$, find $g(-3)$

15) $k(n) = -4n + 5$, find $k(-5)$

16) $w(n) = -3n - 6$, find $w(-4)$

17) $z(n) = 14 - 2n$, find $n(3)$

Perform the indicated operation.

18) $f(x) = 2x + 3$
    $g(x) = x + 4$
    Find $(f - g)(x)$

19) $g(a) = -3a + 2$
    $h(a) = a^2 - 4$
    Find $(h + g)(3)$

Perform the indicated operation.

20) $g(x) = x - 4$
    $f(x) = x + 3$
    Find $(g \cdot f)(2)$

21) $f(x) = x + 2$
    $h(x) = x - 5$
    Find $\left(\frac{f}{h}\right)(-3)$

Using $f(x) = 2x + 5$ and $g(x) = 2x - 3$, find:

22) $g(f(2)) = $ ___

23) $g(f(-2)) = $ ___

24) $f(f(1)) = $ ___

25) $f(f(-1)) = $ ___

26) $f(g(4)) = $ ___

27) $f(g(-3)) = $ ___

Day 10: Answers

1) Mode = the value in the list that appears most often = 9
   Range: the difference of the largest value and smallest value. Then:
   $11 - 8 = 3$

2) Mean: $\frac{sum\ of\ the\ data}{total\ number\ of\ data\ entires} = \frac{6+8+1+4+7+6+10}{7} = \frac{42}{7} = 6$
   Median: is the middle number of a group of numbers arranged in order by size. Write the numbers in order: 1, 4, 6, 7, 8, 10. Then: Median $= \frac{6+7}{2} = 6.5$

3) Bob's rent = $675

   Total expenses → 45% of total expenses = rent →

   Total expenses = rent ÷ 0.45 = $675 ÷ 0.45 = $1,500

4) Bills = 28% × total expenses = 0.28 × $1,500 = $420

5) Car = 14% × total expenses = 0.14 × $1,500 = 210

6) Probability of pulling out a red marble from bag A = $\frac{numbers\ of\ red\ marbles}{total\ of\ marbles} = \frac{9}{6+9} = \frac{9}{15} = 0.6 = 60\%$
   Probability of pulling out a black marble from bag B = $\frac{numbers\ of\ black\ marbles}{total\ of\ marbles} = \frac{6}{6+9} = \frac{6}{15} = 0.4 = 40\%$

7) Jason has 3 choices. Therefore, number of different ways of ordering the events is: $3 \times 2 \times 1 = 6$

8) This is a permutation problem. (order is important) Then: $nP_k = \frac{n!}{(n-k)!} \to n = 8, k = 2 \to \frac{8!}{(8-2)!} = \frac{8!}{6!} = \frac{8 \times 7 \times 6!}{6!} = 56$

9) This is a combination problem (order doesn't matter) $nC_r = \frac{n!}{r!\,(n-r)!} \to$
   $n = 12, r = 9 \to nC_r = \frac{12!}{9!\,(12-9)!} = \frac{12!}{9!\,3!} = \frac{12 \times 11 \times 10 \times 9!}{3 \times 2 \times 1 \times 9!} = \frac{1,320}{6} = 220$

10) $g(n) = 3n + 7 \to g(3) = 3(3) + 7 = 9 + 7 = 16$

11) $h(x) = 4n - 7 \to h(5) = 4(5) - 7 = 20 - 7 = 13$

12) $y(n) = 12 - 3n \to y(8) = 12 - 3(8) = 12 - 24 = -12$

13) $b(n) = -10 - 5n \to b(8) = -10 - 5(8) = -10 - 40 = -50$

14) $g(x) = -7x + 6 \to g(-3) = -7(-3) + 6 = 21 + 6 = 27$

15) $k(n) = -4n + 5 \to k(-5) = -4(-5) + 5 = 20 + 5 = 25$

16) $w(n) = -3n - 6 \rightarrow w(-4) = -3(-4) - 6 = 12 - 6 = 6$

17) $z(n) = 14 - 2n \rightarrow n(3) = 14 - 2(3) = 14 - 6 = 8$

18) $(f - g)(x) = f(x) - g(x) = 2x + 3 - (x + 4) = 2x + 3 - x - 4 = x - 1$

19) $(h + g)(3) = h(a) + g(a) = a^2 - 4 - 3a + 2 = a^2 - 3a - 2 = (3)^2 - 3(3) - 2 = 9 - 9 - 2 = -2$

20) $(g.f)(2) = f(x) \times g(x) = (x - 4) \times (x + 3) = (x \times x) + (x \times 3) + (-4 \times x) + (-4 \times 3) = x^2 + 3x - 4x - 12 = x^2 - x - 12 = 2^2 - 2 - 12 = 4 - 2 - 12 = -10$

21) $\left(\frac{f}{h}\right)(-3) = \frac{f(x)}{h(x)} = \frac{x+2}{x-5} = \frac{(-3)+2}{(-3)-5} = \frac{-3+2}{-3-5} = \frac{-1}{-8} = \frac{1}{8}$

22) First, find $f(2)$: $f(x) = 2x + 5 \rightarrow f(2) = 2(2) + 5 = 9$

    Then: $g(f(2)) = g(9)$. Now, find $g(9)$ by substituting x with 9 in $g(x)$ function. $g(f(2)) = g(9) = 2(9) - 3 = 18 - 3 = 15$

23) First, find $f(-2)$: $f(x) = 2x + 5 \rightarrow f(-2) = 2(-2) + 5 = -4 + 5 = 1$

    Then: $g(f(-2)) = g(1)$. Now, find $g(1)$ by substituting x with 1 in $g(x)$ function. $g(f(-2)) = g(1) = 2(1) - 3 = 2 - 3 = -1$

24) First, find $f(1)$: $f(x) = 2x + 5 \rightarrow f(1) = 2(1) + 5 = 2 + 5 = 7$

    Then: $f(f(1)) = f(7)$. Now, find $f(7)$ by substituting x with 7 in $f(x)$ function. $f(f(1)) = f(7) = 2(7) + 5 = 14 + 5 = 19$

25) First, find $f(-1)$: $f(x) = 2x + 5 \rightarrow f(-1) = 2(-1) + 5 = -2 + 5 = 3$

    Then: $f(f(-1)) = f(3)$. Now, find $f(3)$ by substituting x with 3 in $f(x)$ function. $f(g(-1)) = f(3) = 2(3) + 5 = 6 + 5 = 11$

26) First, find $g(4)$: $g(x) = 2x - 3 \rightarrow g(4) = 2(4) - 3 = 8 - 3 = 5$

    Then: $f(g(4)) = f(5)$. Now, find $f(5)$ by substituting x with 5 in $f(x)$ function. $f(g(4)) = f(5) = 2(5) + 5 = 10 + 5 = 15$

27) First, find $g(-3)$: $g(x) = 2x - 3 \rightarrow g(-3) = 2(-3) - 3 = -6 - 3 = -9$

    Then: $f(g(-3)) = f(-9)$.

    Now, find $f(-9)$ by substituting x with -9 in $f(x)$. function. $f(g(-3)) = f(-9) = 2(-9) + 5 = -18 + 5 = -13$

Cracking CBEST Math Test

Embark on your journey to conquering the CBEST Math exam!

Congratulations on reaching a vital milestone in your CBEST Math preparation. With a solid understanding of the concepts under your belt, you're now ready to elevate your study by applying what you've learned. "***Cracking CBEST Math Test***" is designed to bridge the gap between theory and practice, offering you a realistic testing experience.

Dive into the specifics of what this section offers:

CBEST Math Test-Taking Strategies: Learn techniques that can help streamline your problem-solving process and enhance your accuracy under timed conditions.

CBEST Math – Test Day Tips: Equip yourself with actionable advice to maintain calm and focus when it counts.

Now, put your knowledge into action:

CBEST Math Practice Test 1 and 2: Step into the test-taker's shoes with practice tests crafted to mirror the actual exam's format and rigor.

CBEST Math Practice Tests Answer Keys: Verify your solutions and understand your strengths and areas for improvement.

Answers and Explanations for Practice Tests 1 and 2: Benefit from detailed explanations to deepen your comprehension and correct misunderstandings.

Each practice question you work through is an opportunity to refine your skills and build confidence. Embrace this phase as a key part of your journey towards CBEST Math success.

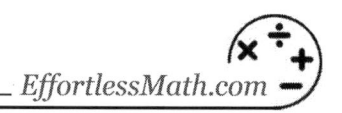

CBEST Math Test-Taking Strategies

Here are some test-taking strategies that you can use to maximize your performance and results on the CBEST Math test.

#1: Use This Approach To Answer Every CBEST Math Question

- Review the question to identify keywords and important information.
- Translate the keywords into math operations so you can solve the problem.
- Review the answer choices. What are the differences between answer choices?
- Draw or label a diagram if needed.
- Try to find patterns.
- Find the right method to answer the question. Use straightforward math, plug in numbers, or test the answer choices (backsolving).
- Double-check your work.

#2: Use Educated Guessing

This approach is applicable to the problems you understand to some degree but cannot solve using straightforward math. In such cases, try to filter out as many answer choices as possible before picking an answer. In cases where you don't have a clue about what a certain problem entails, don't waste any time trying to eliminate answer choices. Just choose one randomly before moving onto the next question.

As you can ascertain, direct solutions are the most optimal approach. Carefully read through the question, determine what the solution is using the math you have learned before, then coordinate the answer with one of the choices available to you. Are you stumped? Make your best guess, then move on.

Don't leave any fields empty! Even if you're unable to work out a problem, strive to answer it. Take a guess if you have to. You will not lose points by getting an answer wrong, though you may gain a point by getting it correct!

#3: Ballpark

A ballpark answer is a rough approximation. When we become overwhelmed by calculations and figures, we end up making silly mistakes. A decimal that is moved by one unit can change an answer from right to wrong, regardless of the number of steps that you went through to get it. That's where ballparking can play a big part.

If you think you know what the correct answer may be (even if it's just a ballpark answer), you'll usually have the ability to eliminate a couple of choices. While answer choices are usually based on the average student error and/or values that are closely tied, you will still be able to weed out choices that are way far afield. Try to find answers that aren't in the proverbial ballpark when you're looking for a wrong answer on a multiple-choice question. This is an optimal approach to eliminating answers to a problem.

#4: Backsolving

All questions on the CBEST Math test will be in multiple-choice format. Many test-takers prefer multiple-choice questions, as at least the answer is right there. You'll typically have five answers to pick from. You simply need to figure out which one is correct. Usually, the best way to go about doing so is "backsolving."

As mentioned earlier, direct solutions are the most optimal approach to answering a question. Carefully read through a problem, calculate a solution, then correspond the answer with one of the choices displayed in front of you. If you can't calculate a solution, your next best approach involves "backsolving."

When backsolving a problem, contrast one of your answer options against the problem you are asked, then see which of them is most relevant. More often than not, answer choices are listed in ascending or descending order. In such cases, try out the choices B or C. If it's not correct, you can go either down or up from there.

#5 : Plugging In Numbers

"Plugging in numbers" is a strategy that can be applied to a wide range of different math problems on the CBEST Math test. This approach is typically used to simplify a challenging question so that it is more understandable. By using the strategy carefully, you can find the answer without too much trouble.

The concept is fairly straightforward–replace unknown variables in a problem with certain values. When selecting a number, consider the following:

- Choose a number that's basic (just not too basic). Generally, you should avoid choosing 1 (or even 0). A decent choice is 2.

- Try not to choose a number that is displayed in the problem.

- Make sure you keep your numbers different if you need to choose at least two of them.

- More often than not, choosing numbers merely lets you filter out some of your answer choices. As such, don't just go with the first choice that gives you the right answer.

- If several answers seem correct, then you'll need to choose another value and try again. This time, though, you'll just need to check choices that haven't been eliminated yet.

- If your question contains fractions, then a potential right answer may involve either an LCD (least common denominator) or an LCD multiple.

- 100 is the number you should choose when you are dealing with problems involving percentages.

CBEST Mathematics Test – Daytime Tips

After practicing and reviewing all the math concepts you've been taught, and taking some CBEST mathematics practice tests, you'll be prepared for test day. Consider the following tips to be extra-ready come test time.

Before Your Test

What to do the night before:

- **Relax!** One day before your test, study lightly or skip studying altogether. You shouldn't attempt to learn something new, either. There are plenty of reasons why studying the evening before a big test can work against you. Put it this way–a marathoner wouldn't go out for a sprint before the day of a big race. Mental marathoners–such as yourself–should not study for any more than one hour 24 hours before a CBEST test. That's because your brain requires some rest to be at its best. The night before your exam, spend some time with family or friends, or read a book.

- **Avoid bright screens** - You'll have to get some good shuteye the night before your test. Bright screens (such as the ones coming from your laptop, TV, or mobile device) should be avoided altogether. Staring at such a screen will keep your brain up, making it hard to drift asleep at a reasonable hour.

- **Make sure your dinner is healthy** - The meal that you have for dinner should be nutritious. Be sure to drink plenty of water as well. Load up on your complex carbohydrates, much like a marathon runner would do. Pasta, rice, and potatoes are ideal options here, as are vegetables and protein sources.

- **Get your bag ready for test day** - The night prior to your test, pack your bag with your stationery, admissions pass, ID, and any other gear that you need. Keep the bag right by your front door.

- **Make plans to reach the testing site** - Before going to sleep, ensure that you understand precisely how you will arrive at the site of the test. If parking is something you'll have to find first, plan for it. If you're dependent on public transit, then review the schedule. You should also make sure that the train/bus/subway/streetcar you use will be running. Find out about road closures as well. If a parent or friend is accompanying you, ensure that they understand what steps they have to take as well.

The Day of the Test

- **Get up reasonably early, but not too early.**

- **Have breakfast** - Breakfast improves your concentration, memory, and mood. As such, make sure the breakfast that you eat in the morning is healthy. The last thing you want to be is distracted by a grumbling tummy. If it's not your own stomach making those noises, another test taker close to you might be instead. Prevent discomfort or embarrassment by consuming a healthy breakfast. Bring a snack with you if you think you'll need it.

- **Follow your daily routine** - Do you watch Good Morning America each morning while getting ready for the day? Don't break your usual habits on the day of the test. Likewise, if coffee isn't something you drink in the morning, then don't take up the habit hours before your test. Routine consistency lets you concentrate on the main objective—doing the best you can on your test.

- **Wear layers** - Dress yourself up in comfortable layers. You should be ready for any kind of internal temperature. If it gets too warm during the test, take a layer off.

- **Get there on time** - The last thing you want to do is get to the test site late. Rather, you should be there 45 minutes prior to the start of the test. Upon your arrival, try not to hang out with anybody who is nervous. Any anxious energy they exhibit shouldn't influence you.

- **Leave the books at home** - No books should be brought to the test site. If you start developing anxiety before the test, books could encourage you to do some last-minute studying, which will only hinder you. Keep the books far away–better yet, leave them at home.

- **Make your voice heard** - If something is off, speak to a proctor. If medical attention is needed or if you'll require anything, consult the proctor prior to the start of the test. Any doubts you have should be clarified. You should be entering the test site with a state of mind that is completely clear.

- **Have faith in yourself** - When you feel confident, you will be able to perform at your best. When you are waiting for the test to begin, envision yourself receiving an outstanding result. Try to see yourself as someone who knows all the answers, no matter what the questions are. A lot of athletes tend to use this technique–particularly before a big competition. Your expectations will be reflected by your performance.

During your test

- **Be calm and breathe deeply** - You need to relax before the test, and some deep breathing will go a long way to help you do that. Be confident and calm. You got this. Everybody feels a little stressed out just before an evaluation of any kind is set to begin. Learn some effective breathing exercises. Spend a minute meditating before the test starts. Filter out any negative thoughts you have. Exhibit confidence when having such thoughts.

- **Concentrate on the test** - Refrain from comparing yourself to anyone else. You shouldn't be distracted by the people near you or random noise. Concentrate exclusively on the test. If you find yourself irritated by surrounding noises, earplugs can be used to block sounds off close to you. Don't forget–the test is going to last several hours if you're taking more than one subject of the test. Some of that time will be dedicated to brief sections. Concentrate on the specific section you are working on during a particular moment. Do not let your mind wander off to upcoming or previous sections.

- **Skip challenging questions** - Optimize your time when taking the test. Lingering on a single question for too long will work against you. If you don't know what the answer is to a certain question, use your best guess, and mark the question so you can review it later on. There is no need to spend time attempting to solve something you aren't sure about. That time would be better served handling the questions you can actually answer well. You will not be penalized for getting the wrong answer on a test like this.

- **Try to answer each question individually** - Focus only on the question you are working on. Use one of the test-taking strategies to solve the problem. If you aren't able to come up with an answer, don't get frustrated. Simply skip that question, then move onto the next one.

- **Don't forget to breathe!** Whenever you notice your mind wandering, your stress levels boosting, or frustration brewing, take a thirty-second break. Shut your eyes, drop your pencil, breathe deeply, and let your shoulders relax. You will end up being more productive when you allow yourself to relax for a moment.

- **Review your answer**. If you still have time at the end of the test, don't waste it. Go back and check over your answers. It is worth going through the test from start to finish to ensure that you didn't make a sloppy mistake somewhere.

- **Optimize your breaks** - When break time comes, use the restroom, have a snack, and reactivate your energy for the subsequent section. Doing some stretches can help stimulate your blood flow.

After your test

- **Take it easy** - You will need to set some time aside to relax and decompress once the test has concluded. There is no need to stress yourself out about what you could've said, or what you may have done wrong. At this point, there's nothing you can do about it. Your energy and time would be better spent on something that will bring you happiness for the remainder of your day.

EffortlessMath.com

Time to Test

Time to refine your Math skill with a practice test

In this book, there are two complete CBEST Math Tests. Take these tests to simulate the test day experience. After you've finished, score your test using the answer keys.

Before You Start

- You'll need a pencil, scratch papers, and a timer to take the test.
- For each question, there are five possible answers. Choose which one is best.
- It's okay to guess. There is no penalty for wrong answers.
- After you've finished the test, review the answer key to see where you went wrong.

Good luck!

CBEST Math Practice Test 1

2024

Total number of questions: 50

Total time (Calculator): 90 Minutes

Calculators are prohibited for the CBEST exam.

1) The mean of 50 test scores was calculated as 90. But it turned out that one of the scores was misread as 94 but it was 69. What is the mean?

    A. 25
    B. 85.2
    C. 87
    D. 89.5
    E. 90

2) Two dice are thrown simultaneously, what is the probability of getting a sum of 5 or 8?

    A. $\frac{1}{3}$
    B. $\frac{1}{4}$
    C. $\frac{1}{16}$
    D. $\frac{1}{36}$
    E. $\frac{11}{36}$

3) In the figure below, what is the value of x?

    A. 45°
    B. 67°
    C. 68°
    D. 135°
    E. 180°

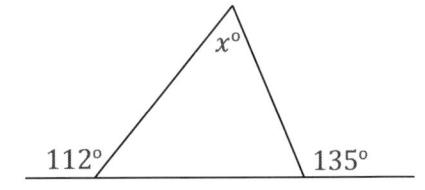

4) What is the value of the expression $6(x - 2y) + (2 - x)^2$ when $x = 3$ and $y = -2$?

    A. -5
    B. 5
    C. 41
    D. 43
    E. 67

5) A swimming pool holds 2,500 cubic feet of water. The swimming pool is 25 feet long and 10 feet wide. How deep is the swimming pool?

    A. 4 $feet$
    B. 6 $feet$
    C. 7 $feet$
    D. 10 $feet$
    E. 25 $feet$

6) Mr. Carlos family are choosing a menu for their reception. They have 2 choices of appetizers, 5 choices of entrees, 4 choices of cake. How many different menu combinations are possible for them to choose?

   A. 12
   B. 20
   C. 32
   D. 40
   E. 60

7) Four one – foot rulers can be split among how many users to leave each with $\frac{1}{3}$ of a ruler?

   A. 4
   B. 6
   C. 12
   D. 24
   E. 48

8) What is the area of a square whose diagonal is 4?

   A. 4
   B. 8
   C. 16
   D. 64
   E. 124

9) In the following right triangle, if the sides AB and BC become twice longer, what will be the ratio of the perimeter of the triangle to its area?

   A. $\frac{1}{2}$
   B. $\frac{1}{5}$
   C. $\frac{3}{2}$
   D. 1
   E. 2

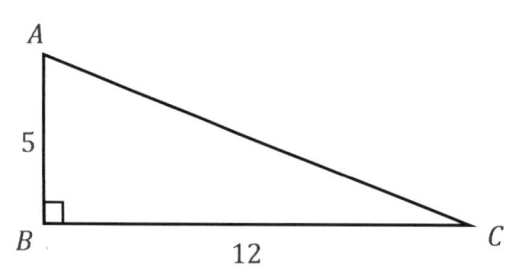

10) The average of five numbers is 26. If a sixth number 42 is added, then, what is the new average? (round your answer to the nearest hundredth).

   A. 25
   B. 26.5
   C. 27
   D. 28.67
   E. 36

11) Mr. Jones saves $2,500 out of his monthly family income of $65,000. What fractional part of his income does he save?

   A. $\frac{1}{11}$
   B. $\frac{1}{15}$
   C. $\frac{1}{26}$
   D. $\frac{2}{15}$
   E. $\frac{3}{25}$

12) A football team had $20,000 to spend on supplies. The team spent $14,000 on new balls. New sport shoes cost $110 each. Which of the following inequalities represent how many new shoes the team can purchase?

   A. $110x + 14,000 \leq 20,000$
   B. $110x + 14,000 \geq 20,000$
   C. $14,000x + 110 \leq 20,000$
   D. $14,000x + 110 \geq 20,000$
   E. $14,000x + 14,000 \geq 20,000$

13) Jason needs an 70% average in his writing class to pass. On his first 4 exams, he earned scores of 68%, 72%, 85%, and 90%. What is the minimum score Jason can earn on his fifth and final test to pass?

   A. 80%
   B. 70%
   C. 68%
   D. 54%
   E. 35%

14) A child grows $1\frac{1}{7}$ inches in $\frac{1}{5}$ of a year. What would be his yearly growth rate in inches per year?

   A. $5\frac{7}{5}$
   B. $5\frac{5}{7}$
   C. $2\frac{1}{7}$
   D. $1\frac{1}{12}$
   E. $\frac{1}{12}$

15) A construction company is building a wall. The company can build 30 cm of the wall per minute. After 40 minutes $\frac{3}{4}$ of the wall is completed. How many meters is the wall?

   A. 4 m
   B. 12 m
   C. 16 m
   D. 30 m
   E. 40 m

16) Kim earned $55 an hour. John earned 10% less than Kim. How much money did John earn in an hour?

   A. $43.50
   B. $45.50
   C. $47.50
   D. $49.50
   E. $51.50

17) Last week 25,000 fans attended a football match. This week three times as many bought tickets, but one sixth of them cancelled their tickets. How many are attending this week?

   A. 48,000
   B. 54,000
   C. 62,500
   D. 75,000
   E. 84,000

18) What is the perimeter of a square that has an area of 49 square inches?

   A. 144 inches
   B. 64 inches
   C. 56 inches
   D. 48 inches
   E. 28 inches

19) If the area of the following rectangular $ABCD$ is 100, and E is the midpoint of AB, what is the area of the shaded part?

   A. 25
   B. 50
   C. 75
   D. 80
   E. 100

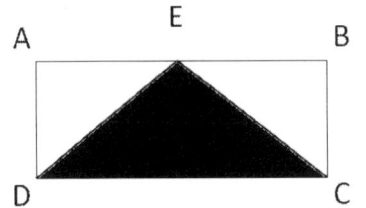

20) Set A contains all integers from 15 to 160, inclusive, and set B contains all integers from 74 to 180, inclusive. How many integers are included in A, but not in B?

   A. 56
   B. 57
   C. 58
   D. 59
   E. 60

21) If the ordered pair $(-4, 5)$ is reflected over the x-axis, what is the new ordered pair?

   A. $(4, 5)$
   B. $(-4, 5)$
   C. $(5, -4)$
   D. $(5, 4)$
   E. $(-4, -5)$

22) What is the ones digit in the divided of the problem below?

$$23 \overline{)29\square} \quad \begin{array}{l} 12 \quad \text{Remainder} = 19 \end{array}$$

   A. 1
   B. 2
   C. 3
   D. 4
   E. 5

23) A taxi driver earns $9 per 1—hour work. If he works 10 hours a day and in 1 hour he uses 2 —liters petrol with price $1 for 1 —liter, how much money does he earn in one day?

   A. $90
   B. $88
   C. $70
   D. $60
   E. $20

24) A cruise line ship left Port A and traveled 50 miles due east and then 120 miles due north. At this point, what is the shortest distance from the cruise to port A?

   A. 70 *miles*
   B. 80 *miles*
   C. 130 *miles*
   D. 150 *miles*
   E. 230 *miles*

25) What is the equivalent temperature of 104°F in Celsius?
$$C = \frac{5}{9}(F - 32)$$

   A. 32
   B. 40
   C. 48
   D. 52
   E. 64

26) Anita's trick—or—treat bag contains 15 pieces of chocolate, 10 suckers, 10 pieces of gum, 25 pieces of licorice. If she randomly pulls a piece of candy from her bag, what is the probability of her pulling out a piece of sucker?

   A. $\frac{1}{3}$
   B. $\frac{1}{4}$
   C. $\frac{1}{6}$
   D. $\frac{1}{12}$
   E. $\frac{1}{24}$

27) What is the missing term in the given sequence?

$$3, 4, 6, 9, 13, 18, 24, \_\_, 39$$

    A. 24
    B. 26
    C. 27
    D. 28
    E. 31

28) The perimeter of a rectangular yard is 72 meters. What is its length if its width is twice its length?

    A. 12 *meters*
    B. 18 *meters*
    C. 20 *meters*
    D. 24 *meters*
    E. 36 *meters*

29) How many positive integers satisfy the inequality $x + 5 < 21$?

    A. 10
    B. 12
    C. 14
    D. 15
    E. 17

30) What is the volume of a box with the following dimensions?
    High = 3 *cm*,   width = 5 *cm*, length = 6 *cm*

    A. 15 cm^3
    B. 60 cm^3
    C. 90 cm^3
    D. 120 cm^3
    E. 240 cm^3

31) In two successive years, the population of a town is increased by 10% and 20%. What percent of the population is increased after two years?

    A. 30%
    B. 32%
    C. 35%
    D. 68%
    E. 70%

32) The sum of six different negative integers is -70. If the smallest of these integers is -15, what is the largest possible value of one of the other five integers?

   A. -15
   B. -14
   C. -10
   D. -5
   E. -1

33) If 20% of a number is 4, what is the number?

   A. 4
   B. 8
   C. 10
   D. 20
   E. 25

34) Jason left a $12.00 tip on a lunch that cost $40.00, approximately what percentage was the tip?

   A. 2.5%
   B. 10%
   C. 15%
   D. 20%
   E. 30%

35) Emily deposits 15% of $160 into a savings account, what is the amount of his deposit?

   A. $10
   B. $16
   C. $20
   D. $24
   E. $30

36) If A is 4 times of B and A is 12, what is the value of B?

   A. 2
   B. 3
   C. 4
   D. 5
   E. 6

37) Jason is 9 miles ahead of Joe running at 6.5 miles per hour and Joe is running at the speed of 8 miles per hour. How long does it take Joe to catch Jason?

   A. 3 hours
   B. 4 hours
   C. 6 hours
   D. 8 hours
   E. 10 hours

38) 44 students took an exam and 11 of them failed. What percent of the students passed the exam?

   A. 20%
   B. 40%
   C. 60%
   D. 75%
   E. 90%

39) In the following figure, MN is 40 cm. How long is ON?

   A. 25 cm
   B. 20 cm
   C. 15 cm
   D. 10 cm
   E. 5 cm

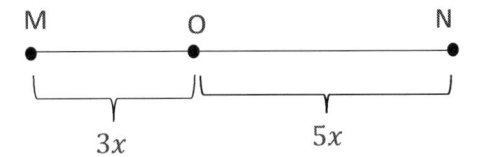

40) The diagonal of a rectangle is 10 inches long and the height of the rectangle is 6 inches. What is the perimeter of the rectangle?

   A. 10 inches
   B. 12 inches
   C. 16 inches
   D. 18 inches
   E. 28 inches

41) The cost, in thousands of dollars, of producing x thousands of textbooks is $C(x) = x^2 + 2x$. The revenue, also in thousands of dollars, is $R(x) = 40x$. Find the profit or loss if 30 textbooks are produced. (Profit = revenue – cost)

   A. $2,160 profit
   B. $2,160 loss
   C. $1,200 loss
   D. $240 profit
   E. $240 loss

42) If angle AOD measures 23°, what is the measure of angle DOC ?

    A. 23°

    B. 46°

    C. 67°

    D. 157°

    E. It cannot be determined from information given.

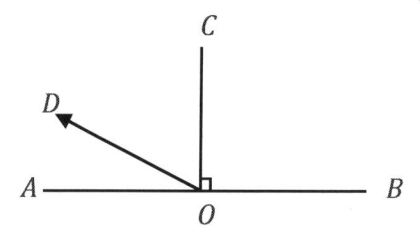

43) A card is drawn at random from a standard 52–card deck, what is the probability that the card is of Hearts? (The deck includes 13 of each suit clubs, diamonds, hearts, and spades)

    A. $\frac{1}{2}$

    B. $\frac{1}{4}$

    C. $\frac{1}{6}$

    D. $\frac{1}{52}$

    E. $\frac{1}{104}$

44) Which of the following shows the numbers in descending order?

$$\frac{1}{3}, 0.68, 67\%, \frac{4}{5}$$

    A. $67\%, 0.68, \frac{1}{3}, \frac{4}{5}$

    B. $67\%, 0.68, \frac{4}{5}, \frac{1}{3}$

    C. $0.68, 67\%, \frac{1}{3}, \frac{4}{5}$

    D. $\frac{1}{3}, 67\%, 0.68, \frac{4}{5}$

    E. $\frac{1}{3}, 67\%, \frac{4}{5}, 0.68$

45) Which of the following is the best estimate for $3{,}689 \times 340$?

    A. 9,000,000

    B. 1,200,000

    C. 900,000

    D. 120,000

    E. 9,000

46) The table below shows a relationship between values of x and y. What is the value of y that is missing?

A. 4.75
B. 5.25
C. 5.5
D. 5.25
E. 5.75

x	y
2	3
2.5	3.75
3	4.5
3.5	...
4	6

47) What is the percent of cars is blue?

A. 18%
B. 20%
C. 23%
D. 26%
E. 33%

Cars of company

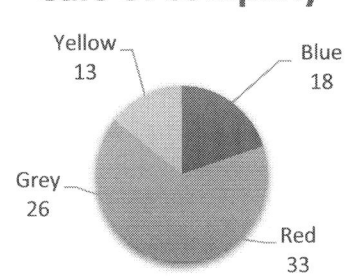

Yellow 13
Blue 18
Grey 26
Red 33

Use the table below to answer the question.

Students	Boys	Girls
Grade 5	24	16
Grade 6	18	27
Grade 7	14	19

48) The table above shows the number of students in a school. What percent of sixth grade students are girls?

A. 40%
B. 42%
C. 55%
D. 60%
E. 62%

Use the chart below to answer the question. The table below shows the number of different color marbles in a bag.

Color	Number of marbles
White	20
Black	30
Beige	40

49) There are also purple marbles in the bag. Which of the following can NOT be the probability of randomly selecting a purple marble from the bag?

A. $\frac{1}{10}$
B. $\frac{1}{4}$
C. $\frac{2}{5}$
D. $\frac{7}{15}$
E. $\frac{9}{15}$

Gender	Under 45	45 or older	total
Male	12	6	18
Female	5	7	12
Total	17	13	30

50) The table above shows the distribution of age and gender for 30 employees in a company. If one employee is selected at random, what is the probability that the employee selected be either a female under age 45 or a male age 45 or older?

A. $\frac{5}{6}$
B. $\frac{5}{30}$
C. $\frac{6}{30}$
D. $\frac{11}{30}$
E. $\frac{45}{30}$

End of CBEST Math Practice Test 1.

CBEST Math Practice Test 2

2024

Total number of questions: 50

Total time (Calculator): 90 Minutes

Calculators are prohibited for the CBEST exam.

1) If the area of the following trapezoid is 126 *cm*, what is the perimeter of the trapezoid? (Figure not drawn to scale.)

   A. 32
   B. 42
   C. 46
   D. 56
   E. 64

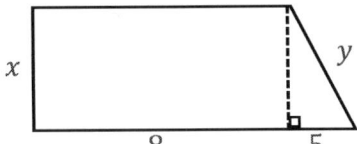

2) If 5 inches on a map represents an actual distance of 100 feet, then, what actual distance does 18 inches on the map represent?

   A. 18 feet
   B. 20 feet
   C. 100 feet
   D. 250 feet
   E. 360 feet

3) Which of the following lists shows the fractions in order from least to greatest?

   $$\frac{3}{4}, \frac{2}{7}, \frac{3}{8}, \frac{5}{11}$$

   A. $\frac{3}{8}, \frac{2}{7}, \frac{3}{4}, \frac{5}{11}$
   B. $\frac{3}{8}, \frac{2}{7}, \frac{5}{11}, \frac{3}{4}$
   C. $\frac{2}{7}, \frac{5}{11}, \frac{3}{8}, \frac{3}{4}$
   D. $\frac{2}{7}, \frac{3}{8}, \frac{5}{11}, \frac{3}{4}$
   E. $\frac{5}{11}, \frac{3}{4}, \frac{3}{8}, \frac{2}{7}$

4) What is the value of 6^4?

   A. 6
   B. 24
   C. 36
   D. 216
   E. 1,296

5) How many $\frac{1}{5}$ pound paperback books together weigh 50 pounds?

   A. 25
   B. 50
   C. 150
   D. 200
   E. 250

6) What is the volume of the following square pyramid?

   A. $100\ m^3$
   B. $120\ m^3$
   C. $144\ m^3$
   D. $480\ m^3$
   E. $1,440\ m^3$

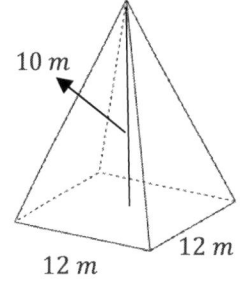

7) What is the value of x in the following equation?

   $$\frac{2}{3}x + \frac{1}{6} = \frac{1}{3}$$

   A. 6
   B. $\frac{1}{2}$
   C. $\frac{1}{3}$
   D. $\frac{1}{4}$
   E. $\frac{1}{12}$

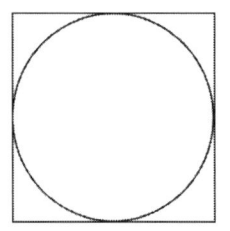

8) In the following shape, the area of the circle is 16π. What is the area of the square?

   A. 4
   B. 8
   C. 16
   D. 32
   E. 64

9) List A consists of the numbers $\{1, 3, 8, 10, 15\}$, and list B consists of the numbers $\{4, 6, 12, 14, 17\}$. If the two lists are combined, what is the median of the combined list?

   A. 9
   B. 10
   C. 12
   D. 15
   E. 17

10) What's the area of the non-shaded part of the following figure?

   A. 236
   B. 192
   C. 152
   D. 42
   E. 40

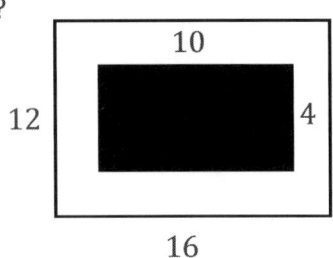

11) In the triangle below, if the measure of angle A is 37 degrees, then what is the value of y? (figure is NOT drawn to scale)

   A. 37
   B. 62
   C. 70
   D. 78
   E. 86

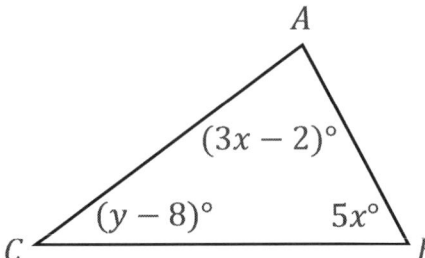

12) There are only red and blue cards in a box. The probability of choosing a red card in the box at random is one third. If there are 246 blue cards, how many cards are in the box?

   A. 123
   B. 246
   C. 308
   D. 328
   E. 369

13) In the diagram below, circle A represents the set of all odd numbers, circle B represents the set of all negative numbers, and circle C represents the set of all multiples of 5. Which number could be replaced with y?

   A. 0
   B. 5
   C. −5
   D. 10
   E. −10

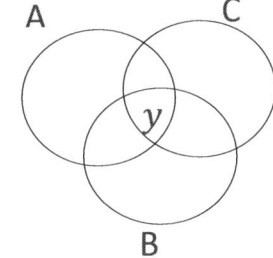

14) Which of the following is an obtuse angle?

   A. 116°
   B. 80°
   C. 68°
   D. 25°
   E. 15°

15) A basket contains 20 balls and the average weight of each of these balls is 25 g. The five heaviest balls have an average weight of 40 g each. If we remove the five heaviest balls from the basket, what is the average weight of the remaining balls?

   A. 10 g
   B. 20 g
   C. 30 g
   D. 35 g
   E. 40 g

16) In a stadium the ratio of home fans to visiting fans in a crowd is $5:7$. Which of the following could be the total number of fans in the stadium?

   A. 12,324
   B. 42,326
   C. 44,566
   D. 66,812
   E. 69,752

17) A bread recipe calls for $2\frac{2}{3}$ cups of flour. If you only have $1\frac{5}{6}$ cups, how much more flour is needed?

   A. 1
   B. 2
   C. $\frac{1}{2}$
   D. $\frac{5}{6}$
   E. $\frac{11}{6}$

18) If $x = \frac{1}{3}$ and $y = \frac{9}{21}$, then which is equal to $\frac{1}{x} \div \frac{y}{3}$?

   A. $\frac{1}{7}$
   B. $\frac{1}{3}$
   C. $\frac{2}{3}$
   D. $\frac{1}{21}$
   E. 21

19) Ella (E) is 4 years older than her friend Ava (A) who is 3 years younger than her sister Sofia (S). If E, A and S denote their ages, which one of the following represents the given information?

A. $\begin{cases} E = A + 4 \\ S = A - 3 \end{cases}$

B. $\begin{cases} E = A + 4 \\ A = S + 3 \end{cases}$

C. $\begin{cases} A = E + 4 \\ S = A - 3 \end{cases}$

D. $\begin{cases} E = A + 4 \\ A = S - 3 \end{cases}$

E. $\begin{cases} E = A + 3 \\ S = A + 4 \end{cases}$

20) If Jim adds 100 stamps to his current stamp collection, the total number of stamps will be equal to $\frac{6}{5}$ the current number of stamps. If Jim adds 50% more stamps to the current collection, how many stamps will be in the collection?

A. 150

B. 300

C. 500

D. 600

E. 750

21) The sum of 8 numbers is greater than 240 and less than 320. Which of the following could be the average (arithmetic mean) of the numbers?

A. 25

B. 30

C. 35

D. 40

E. 45

22) In the following figure, point Q lies on line n, what is the value of y if $x = 35$?

A. 21

B. 25

C. 35

D. 40

E. 50

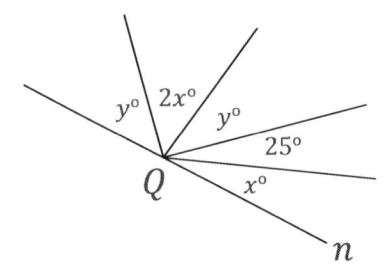

23) The length of a rectangle is 3 meters greater than 4 times its width. The perimeter of the rectangle is 36 meters. What is the area of the rectangle in meters?

   A. 15
   B. 35
   C. 45
   D. 55
   E. 65

24) What is the value of x?

   A. 38
   B. 45
   C. 75
   D. 83
   E. 135

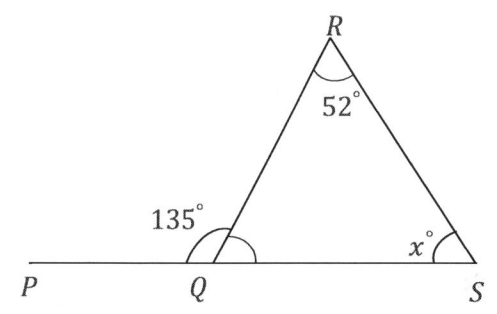

25) Find the value of x in the following diagram. (there are 2 supplementary angles in the diagram)

   A. 27
   B. 32
   C. 37
   D. 45
   E. 47

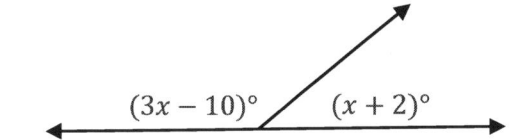

26) What is the area of the shaded region if the diameter of the bigger circle is 12 inches and the diameter of the smaller circle is 8 inches?

   A. $16\pi\ in^2$
   B. $20\pi\ in^2$
   C. $36\pi\ in^2$
   D. $48\pi\ in^2$
   E. $80\pi\ in^2$

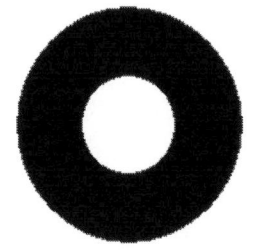

27) Which of the following expressions is equivalent to $\frac{a+b}{2}$?

   A. $\frac{ab}{2}$
   B. $\frac{a}{2} \times \frac{b}{2}$
   C. $\frac{1}{2} \times a \times b$
   D. $\frac{a}{2} + \frac{b}{2}$
   E. $\frac{1}{2}(a \times b)$

28) What is 2.5% of 1,200?

   A. 900
   B. 600
   C. 300
   D. 60
   E. 30

29) If x is a real number, and if $x^3 + 18 = 130$, then x lies between which two consecutive integers?

   A. 1 and 2
   B. 2 and 3
   C. 3 and 4
   D. 4 and 5
   E. 5 and 6

30) Jack types 72 words per minute. How many words does he type in 15 seconds?

   A. 14
   B. 18
   C. 20
   D. 22
   E. 24

31) Jack earns $616 for his first 44 hours of work in a week and is then paid 1.5 times his regular hourly rate for any additional hours. This week, Jack needs $826 to pay his rent, bills and other expenses. How many hours must he work to make enough money in this week?

   A. 27
   B. 32
   C. 44
   D. 54
   E. 68

32) Which of the following is the same as: 0.000,000,000,000,042,121?

   A. 4.2121×10^{14}

   B. 4.2121×10^{13}

   C. 42.121×10^{-10}

   D. 42.121×10^{-13}

   E. 4.2121×10^{-14}

33) Which of the following is the largest?

   A. $|4 - 2|$
   B. $|2 - 4|$
   C. $|-2 - 4|$
   D. $|2 - 4| - |4 - 2|$
   E. $|2 - 4| + |4 - 2|$

34) A student gets 85% of a test with 40 questions. How many answers did the student solve correctly?

   A. 15
   B. 24
   C. 26
   D. 34
   E. 36

35) To buy a new computer, Emma borrowed $2,500 at 8% interest for 6 years. How much interest did she pay?

   A. $150

   B. $1,200

   C. $1,500

   D. $2,400

   E. $2,500

36) Integer x is evenly divisible by 4. Which expression below is also evenly divisible by 4?

   A. $x + 1$
   B. $2x + 1$
   C. $2x + 4$
   D. $3x + 2$
   E. $4x + 1$

Use the figure below to answer the question.

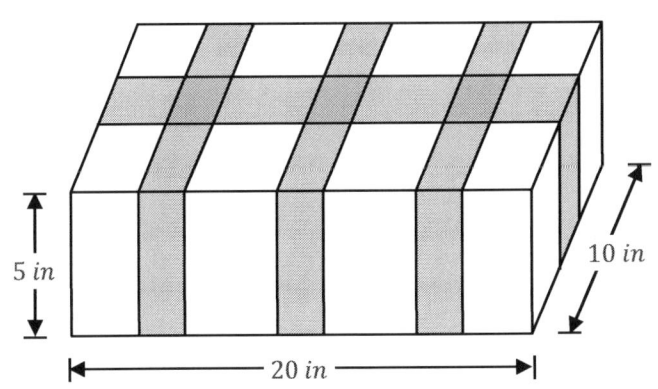

37) Ribbon is wrapped around the box, if 10% is added to the length of the ribbon for the bow tie, what is the length of the ribbon?

   A. $100\ in$
   B. $110\ in$
   C. $140\ in$
   D. $150\ in$
   E. $154\ in$

38) Sara has a box containing 5 blue balls, 8 red balls, and 3 green balls. If she removes one ball at random, what is the probability that it will not be blue?

   A. $\frac{1}{8}$
   B. $\frac{5}{16}$
   C. $\frac{5}{11}$
   D. $\frac{10}{11}$
   E. $\frac{11}{16}$

39) Jack rides 160 kilometers in 1 hour 20 minutes. At that rate, how many meters does he ride per minute?

   A. 1,000 meters
   B. 1,500 meters
   C. 1,600 meters
   D. 2,000 meters
   E. 2,500 meters

40) The sum of two consecutive integer is −13. If 2 is added to the smaller integer and 3 is subtract from the larger integer, what is the product of the two resulting integers?

   A. 5
   B. 9
   C. 18
   D. 28
   E. 45

41) If n is an even integer that is less than −3.34, what is the greatest possible value of n?

   A. −1
   B. −2
   C. −3
   D. −4
   E. −5

42) Find the value of x? $(2)^3 + (-3)^2 + 2x - 6 = 11$

   A. 0
   B. 2
   C. 4
   D. 6
   E. 8

43) A ladder leans against a wall forming a 60° angle between the ground and the ladder. If the bottom of the ladder is 30 feet away from the wall, how long is the ladder?

   A. 30 feet
   B. 40 feet
   C. 50 feet
   D. 60 feet
   E. 120 feet

Use the figure below answer the question.

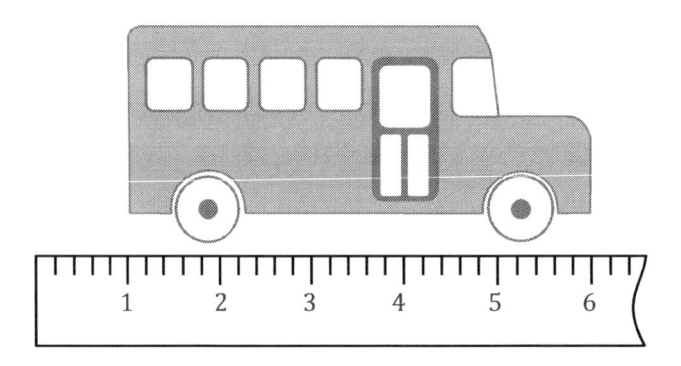

44) In the scale of the diagram is 1 unit = 160 centimeters, what is the actual size of the bus?

   A. 80 centimeters
   B. 800 centimers
   C. 96 centimeters
   D. 960 centimeters
   E. 9,600 centimeters

45) The table below displays a relationship between values of x and y. Which of the following expressions describe this relationship?

   A. $y = \frac{2}{3}x$
   B. $y = \frac{3}{2}x$
   C. $y = 2(x-1)$
   D. $y = 3x - 1$
   E. $y = 4x - 2$

x	y
2	5
4	11
6	17
8	23
10	29

Questions 46 to 48 are based on the following data

Types of air pollutions in 10 cities of a country

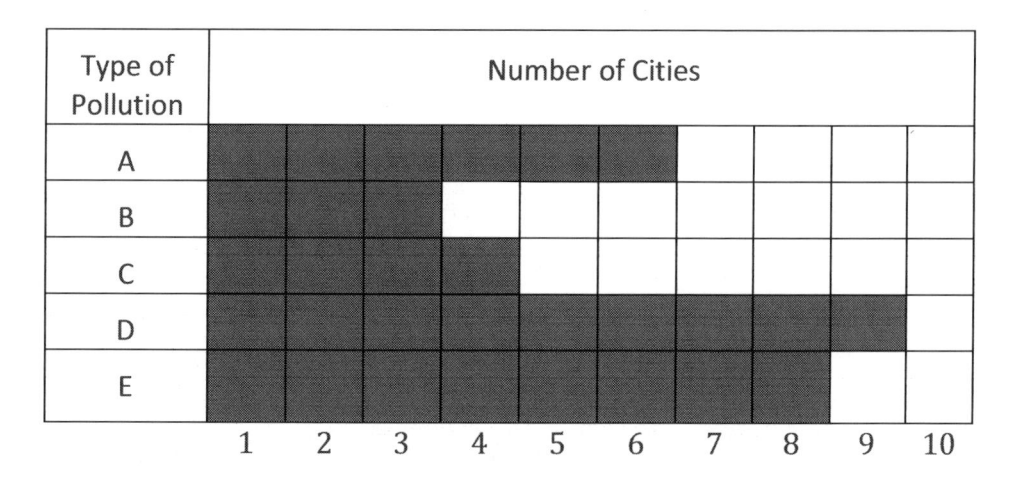

Type of Pollution	Number of Cities
A	6
B	4
C	4
D	8
E	8

  1 2 3 4 5 6 7 8 9 10

46) If a is the mean (average) of the number of cities in each pollution type category, b is the mode, and c is the median of the number of cities in each pollution type category, then which of the following must be true?

A. $a < b < c$
B. $b < a < c$
C. $b < c < b$
D. $a = c$
E. $b < c = a$

47) What percent of cities are in the type of pollution A, C, and D respectively?

A. 60%, 40%, 90%
B. 40%, 90%, 60%
C. 40%, 60%, 90%
D. 30%, 40%, 90%
E. 30%, 40%, 60%

48) How many cities should be added to type of pollutions B until the ratio of cities in type of pollution B to cities in type of pollution E will be 0.625?

A. 2
B. 3
C. 4
D. 5
E. 6

Use the table below to answer the question.

Appliances	Price
Sofa	$365.76
Washing machine	$289.55
Oven	$378.45
TV	$289.99
Refrigerators	$1,459
Dishwasher	?

49) Emily has bought appliances for her new home. The total cost of her purchase is $3,332.49. What is the missing price?

   A. $450.74
   B. $549.47
   C. $549.74
   D. $640.64
   E. $748.74

50) The following graph shows the mark of seven students in mathematics. What is the mean (average) of the marks?

   A. 15
   B. 14.5
   C. 14
   D. 13.5
   E. 13

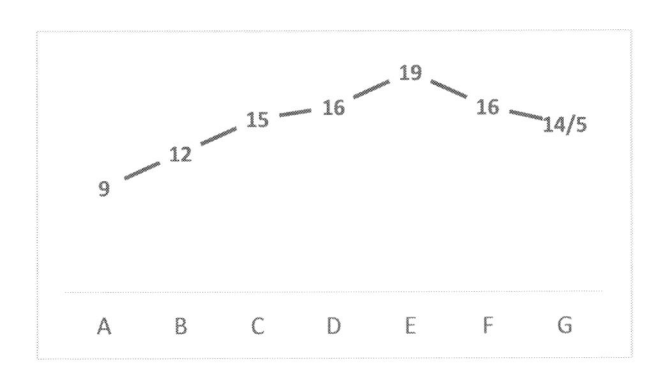

Graph values: A=9, B=12, C=15, D=16, E=19, F=16, G=14/5

End of CBEST Math Practice Test 2.

CBEST Math Practice Tests Answer Keys

Now, it's time to review your results to see where you went wrong and what areas you need to improve.

CBEST Practice Test 1						**CBEST Practice Test 2**					
1	D	21	E	41	D	1	C	21	C	41	D
2	B	22	E	42	C	2	E	22	B	42	A
3	B	23	C	43	B	3	D	23	C	43	D
4	D	24	C	44	D	4	E	24	D	44	B
5	D	25	B	45	B	5	E	25	E	45	D
6	D	26	C	46	D	6	D	26	B	46	D
7	C	27	E	47	B	7	D	27	D	47	A
8	B	28	A	48	D	8	E	28	E	48	A
9	A	29	D	49	D	9	A	29	D	49	C
10	D	30	C	50	D	10	C	30	B	50	B
11	C	31	B			11	E	31	D		
12	A	32	D			12	E	32	E		
13	E	33	D			13	C	33	C		
14	B	34	E			14	A	34	D		
15	C	35	D			15	B	35	B		
16	D	36	B			16	A	36	C		
17	C	37	C			17	D	37	E		
18	E	38	D			18	E	38	E		
19	B	39	A			19	D	39	D		
20	D	40	E			20	E	40	E		

CBEST Math Practice Tests Answers and Explanations

CBEST Math Practice Test 1 Answers and Explanations

1) **Choice D is correct**

$average\ (mean) = \frac{sum\ of\ terms}{number\ of\ terms} \Rightarrow 90 = \frac{sum\ of\ terms}{50} \Rightarrow sum = 90 \times 50 = 4,500$

The difference of 94 and 69 is 25. Therefore, 25 should be subtracted from the sum.

$4,500 - 25 = 4,475, mean = \frac{sum\ of\ terms}{number\ of\ terms} \Rightarrow mean = \frac{4,475}{50} = 89.5$

2) **Choice B is correct**

For sum of 5: (1 & 4) and (4 & 1), (2 & 3) and (3 & 2), therefore we have 4 options.

For sum of 8: (5 & 3) and (3 & 5), (4 & 4) and (2 & 6), and (6 & 2), we have 5 options. To get a sum of 5 or 8 for two dice: $4 + 5 = 9$. Since, we have $6 \times 6 = 36$ total number of options, the probability of getting a sum of 5 and 8 is 9 out of 36 or $\frac{9}{36} = \frac{1}{4}$

3) **Choice B is correct**

First, find the angles α and β. Angles 112 and α are supplementary. Then:

$a + 112 = 180 \to \alpha = 180° - 112° = 68°$

Angles 135 and β are also supplementary. $\beta = 180° - 135° = 45°$

The sum of all angles in a triangle is 180 degrees. Then:

$x + \alpha + \beta = 180° \to x = 180° - 68° - 45° = 67°$

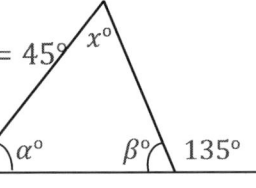

4) **Choice D is correct**

Plug in the value of x and y. $x = 3$ and $y = -2$.

$6(x - 2y) + (2 - x)^2 = 6(3 - 2(-2)) + (2 - 3)^2 = 6(3 + 4) + (-1)^2 = 42 + 1 = 43$

5) **Choice D is correct**

Use formula of rectangle prism volume. $V = (length)(width)(height) \Rightarrow$

$2,500 = (25)(10)(height) \Rightarrow height = 2,500 \div 250 = 10\ feet$

6) Choice D is correct

To find the number of possible outfit combinations, multiply number of options for each factor: $2 \times 5 \times 4 = 40$

7) Choice C is correct

$4 \div \dfrac{1}{3} = 12$

8) Choice B is correct

The diagonal of the square is 4. Let x be the side. Use Pythagorean Theorem: $a^2 + b^2 = c^2$

$x^2 + x^2 = 4^2 \Rightarrow 2x^2 = 4^2 \Rightarrow 2x^2 = 16 \Rightarrow x^2 = 8 \Rightarrow x = \sqrt{8}$

The area of the square is: $\sqrt{8} \times \sqrt{8} = 8$

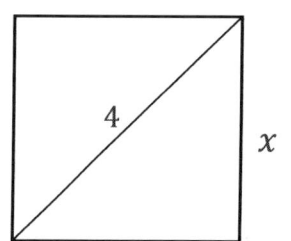

9) Choice A is correct

$AB = 5$ and $BC = 12$, $AC = \sqrt{12^2 + 5^2} = \sqrt{144 + 25} = \sqrt{169} = 13$

Perimeter $= 5 + 12 + 13 = 30$, Area $= \dfrac{5 \times 12}{2} = 5 \times 6 = 30$

In this case, the ratio of the perimeter of the triangle to its area is: $\dfrac{30}{30} = 1$

If the sides AB and BC become twice longer, then: $AB = 10$ And $BC = 24$

$AC = \sqrt{24^2 + 10^2} = \sqrt{576 + 100} = \sqrt{676} = 26$

Perimeter $= 26 + 24 + 10 = 60$, Area $= \dfrac{10 \times 24}{2} = 10 \times 12 = 120$

In this case the ratio of the perimeter of the triangle to its area is: $\dfrac{60}{120} = \dfrac{1}{2}$

10) Choice D is correct

Solve for the sum of five numbers.

$average = \dfrac{sum\ of\ terms}{number\ of\ terms} \Rightarrow 26 = \dfrac{sum\ of\ 5\ numbers}{5} \Rightarrow sum\ of\ 5\ numbers = 26 \times 5 = 130$

The sum of 5 numbers is 130. If a sixth number 42 is added, then the sum of 6 numbers is $130 + 42 = 172 \Rightarrow average = \dfrac{sum\ of\ terms}{number\ of\ terms} = \dfrac{172}{6} = 28.67$

11) Choice C is correct

2,500 out of 65,000 equals to $\frac{2,500}{65,000} = \frac{25}{650} = \frac{1}{26}$

12) Choice A is correct

Let x be the number of shoes the team can purchase. Therefore, the team can purchase $110x$. The team had $20,000 and spent $14,000. Now the team can spend on new shoes $6,000 at most. Now, write the inequality: $110x + 14,000 \leq 20,000$

13) Choice E is correct

Jason needs an 70% average to pass for five exams. Therefore, the sum of 5 exams must be at least $5 \times 70 = 350$. The sum of 4 exams is: $68 + 72 + 85 + 90 = 315$

The minimum score Jason can earn on his fifth and final test to pass is:

$350 - 315 = 35$

14) Choice B is correct

Set up a proportion to solve.

$\frac{1\frac{1}{7} in}{\frac{1}{5} yr} = \frac{x \ in}{1 \ yr} \to 1\frac{1}{7} = \frac{1}{5}x \to \frac{8}{7} = \frac{1}{5}x \to \left(\frac{5}{1}\right)\left(\frac{8}{7}\right) = x \to x = \frac{40}{7} \to x = 5\frac{5}{7}$

15) Choice C is correct

The rate of construction company $= \frac{30 \ cm}{1 \ min} = 30 \frac{cm}{min}$

The height of the wall after 40 minutes $= \frac{30 \ cm}{1 \ min} \times 40 \ min = 1,200 \ cm$

Let x be the height of wall, then $\frac{3}{4}x = 1,200 \ cm \to x = \frac{4 \times 1,200}{3} \to x = 1,600 \ cm = 16 \ m$

16) Choice D is correct

If Kim's earning = 100%, then, John's earning is 90% of Kim's earning. Then:

$0.90 \times 55 = 49.50$

17) Choice C is correct

Three times of 25,000 is 75,000. One sixth of them cancelled their tickets. One sixth of 75,000 equals 12,500 ($\frac{1}{6} \times 75,000 = 12,500$). 62,500 (75,000 − 12,500 = 62,500) fans are attending this week.

18) Choice E is correct

The area of the square is 49 inches. Therefore, the side of the square is square root of the area: $\sqrt{49} = 7$ inches.

Four times the side of the square is the perimeter: $4 \times 7 = 28$ *inches*

19) Choice B is correct

Since, E is the midpoint of AB, then the area of all triangles DAE, DEF, CFE and CBE are equal. Let x be the area of one of the triangles, Then: $4x = 100 \rightarrow x = 25$

The area of $DEC = 2x = 2(25) = 50$

20) Choice D is correct

The integers that are included in Set A but not in Set B are 15 through 73. (Note that 74 is included in Set B). To calculate the number of integers between 15 and 73, inclusive, subtract the two endpoints and add 1. (One must be added because the endpoints are both counted in the total) $73 - 15 + 1 = 59$

21) Choice E is correct

Since the ordered pair is reflected over the x-axis, then, the value of x of the point doesn't change and the sign of y changes. $(-4, 5) \Rightarrow (-4, -5)$

22) Choice E is correct

To find the answer, multiply 12 by 23 and add the result to 19:

$(12 \times 23) + 19 = 295$

23) Choice C is correct

$\$9 \times 10 = \$90$, Petrol use: $10 \times 2 = 20$ liters

Petrol cost: $20 \times \$1 = \$20$, Money earned: $\$90 - \$20 = \$70$

24) Choice C is correct

Use the information provided in the question to draw the shape.

Use Pythagorean Theorem: $a^2 + b^2 = c^2$

$50^2 + 120^2 = c^2 \Rightarrow 2{,}500 + 14{,}400 = c^2 \Rightarrow c^2 = 16{,}900 \Rightarrow$

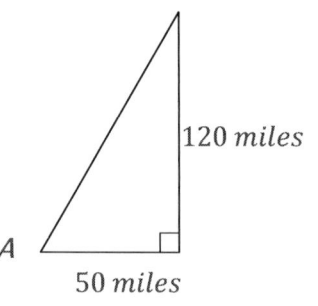

$c = 130$ miles

25) Choice B is correct

Plug in 104 for F and then solve for C.

$C = \dfrac{5}{9}(F - 32) \Rightarrow C = \dfrac{5}{9}(104 - 32) \Rightarrow C = \dfrac{5}{9}(72) = 40$

26) Choice C is correct

$Probability = \dfrac{number\ of\ desired\ outcomes}{number\ of\ total\ outcomes} = \dfrac{10}{15 + 10 + 10 + 25} = \dfrac{10}{60} = \dfrac{1}{6}$

27) Choice E is correct

Find the difference of each pairs of numbers: 3, 4, 6, 9, 13, 18, 24, ___, 39

The difference of 3 and 4 is 1, 4 and 6 is 2, 6 and 9 is 3, 9 and 13 is 4, 13 and 18 is 5, 18 and 24 is 6, 24 and next number should be 7. The number is $24 + 7 = 31$

28) Choice A is correct

The width of the rectangle is twice its length. Let x be the length. Then, $width = 2x$

Perimeter of the rectangle is $2(width + length) = 2(2x + x) = 72 \Rightarrow 6x = 72 \Rightarrow$

$x = 12$. Length of the rectangle is 12 meters.

29) Choice D is correct

First, simplify the inequality: $x + 5 < 21 \rightarrow x < 16$

The positive integers that satisfy the inequality are 1, 2, 3, ..., 14, 15. (We cannot include 16 because x must be less than 16) 15 positive integers satisfy this inequality.

30) Choice C is correct

$Volume\ of\ a\ box = length \times width \times height = 3 \times 5 \times 6 = 90$

31) Choice B is correct

The population is increased by 10% and 20%. 10% increase changes the population to 110% of original population. For the second increase, multiply the result by 120%.

$(1.10) \times (1.20) = 1.32 = 132\%$.

32 percent of the population is increased after two years.

32) Choice D is correct

The smallest number is -15. To find the largest possible value of one of the other five integers, we need to choose the smallest possible integers for four of them. Let x be the largest number. Then: $-70 = (-15) + (-14) + (-13) + (-12) + (-11) + x \rightarrow -70 = -65 + x \rightarrow x = -70 + 65 = -5$

33) Choice D is correct

If 20% of a number is 4, what is the number: 20% of $x = 4 \Rightarrow 0.20x = 4 \Rightarrow$
$x = 4 \div 0.20 = 20$

34) Choice E is correct

$12 is what percent of $40? $\rightarrow 12 \div 40 = 0.30 = 30\%$

35) Choice D is correct

15% of $160 is $0.15 \times 160 = 24$

36) Choice B is correct

A is 4 times of B, then: $A = 4B \Rightarrow (A = 12) 12 = 4 \times B \Rightarrow B = 12 \div 4 = 3$

37) Choice C is correct

The distance between Jason and Joe is 9 *miles*. Jason running at 6.5 *miles per hour* and Joe is running at the speed of 8 *miles per hour*. Therefore, every hour the distance is 1.5 *miles* less. $9 \div 1.5 = 6\ hours$

38) Choice D is correct

The failing rate is 11 out of $44 = \frac{11}{44}$, Change the fraction to percent: $\frac{11}{44} \times 100\% = 25\%$. 25 percent of students failed. Therefore, 75 percent of students passed the exam.

39) Choice A is correct

The length of MN is equal to: $3x + 5x = 8x$. Then: $8x = 40 \rightarrow x = \frac{40}{8} = 5$

The length of ON is equal to: $5x = 5 \times 5 = 25 \, cm$

40) Choice E is correct

Let x be the width of the rectangle. Use Pythagorean Theorem:

$a^2 + b^2 = c^2$

$x^2 + 6^2 = 10^2 \Rightarrow x^2 + 36 = 100 \Rightarrow x^2 = 100 - 36 \Rightarrow x^2 = 64 \Rightarrow x = 8$

Perimeter of the rectangle $= 2(length + width) = 2(8 + 6) = 2(14) = 28$

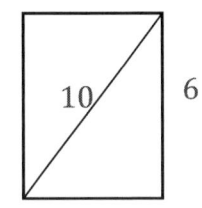

41) Choice D is correct

Plug in the value of $x = 30$ into both equations. Then:

$C(x) = x^2 + 2x = (30)^2 + 2(30) = 900 + 60 = 960$.

$R(x) = 40x = 40 \times 30 = 1,200 \rightarrow 1,200 - 960 = 240$

The profit of producing 30 textbooks is $240.

42) Choice C is correct

Angles AOD and DOC are complementary angles. Therefore, their sum is 90 degrees. Then:

$DOC + AOD = 90° \rightarrow DOC = 90 - AOD \rightarrow$

$DOC = 90 - 23 = 67$

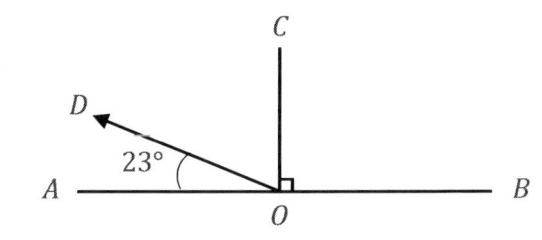

43) Choice B is correct

The probability of choosing a Hearts is: $\frac{13}{52} = \frac{1}{4}$

44) Choice D is correct

Change the numbers to decimal and then compare.

$\frac{1}{3} = 0.\overline{3}$, 0.68, $67\% = 0.67$, $\frac{4}{5} = 0.80$

Therefore $\frac{1}{3} < 67\% < 0.68 < \frac{4}{5}$

45) Choice B is correct

Round both numbers, then multiply: $3,689 \times 340 \rightarrow 4,000 \times 300 = 1,200,000$

46) Choice D is correct

First, find the relationship between y and x. Based on the values provided in the table, the relationship between y and x is: $y = 1.5x$. Now, find the missing y by substituting the value of x in the equation. Then: $y = 1.5x \to y = 1.5 \times 3.5 = 5.25$

47) Choice B is correct

$Percent = \frac{part}{whole} \times 100 = \frac{18}{18+33+26+13} \times 100 = \frac{18}{90} \times 100 = 20\%$

48) Choice D is correct

Number of sixth grade of students $= 18 + 27 = 45$

Number of sixth grade girls students $= 27$

$Percent = \frac{part}{whole} \times 100 \to Percent = \frac{27}{45} \times 100 = 60\%$

49) Choice D is correct

Let x be the number of purple marbles. Let's review the choices provided:

A. $\frac{1}{10}$, if the probability of choosing a purple marble is one out of ten, then:

$$Probability = \frac{number\ of\ desired\ outcomes}{number\ of\ total\ outcomes} = \frac{x}{20+30+40+x} = \frac{1}{10}$$

Use cross multiplication and solve for x. $10x = 90 + x \to 9x = 90 \to x = 10$

Since the number of purple marbles can be 10, then, choice A can be the probability of randomly selecting a purple marble from the bag. Use the same method for other choices.

B. $\frac{1}{4} \to \frac{x}{20+30+40+x} = \frac{1}{4} \to 4x = 90 + x \to 3x = 90 \to x = 30$

C. $\frac{2}{5} \to \frac{x}{20+30+40+x} = \frac{2}{5} \to 5x = 180 + 2x \to 3x = 180 \to x = 60$

D. $\frac{7}{15} \to \frac{x}{20+30+40+x} = \frac{7}{15} \to 15x = 630 + 7x \to 8x = 630 \to x = 78.75$

E. $\frac{9}{15} \to \frac{x}{20+30+40+x} = \frac{9}{15} \to 15x = 810 + 9x \to 6x = 810 \to x = 135$

Number of purple marbles cannot be a decimal. Therefore, Choice D can NOT be the probability of randomly selecting a purple marble from the bag.

50) Choice D is correct

Of the 30 employees, there are 5 females under age 45 and 6 males age 45 or older. Therefore, the probability that the person selected will be either a female under age 45 or a male age 45 or older is: $\frac{5}{30} + \frac{6}{30} = \frac{11}{30}$

CBEST Math Practice Test 2 Answers and Explanations

1) Choice C is correct

The area of the trapezoid is:

$Area = \frac{1}{2}h(b_1 + b_2) \rightarrow 126 = \frac{1}{2}(x)(13 + 8) \rightarrow 126 = 10.5x \rightarrow x = 12$

$y = \sqrt{5^2 + 12^2} = \sqrt{25 + 144} = \sqrt{169} = 13$

The perimeter of the trapezoid is: $12 + 13 + 8 + 13 = 46$

2) Choice E is correct

First calculate the number of feet that 1 inch represents: $100\ ft \div 5\ in = 20\ ft/in$

Then multiply this by the total number of inches: $18\ in \times 20\ ft/in = 360\ ft$

3) Choice D is correct

Let's compare each fraction: $\frac{2}{7} < \frac{3}{8} < \frac{5}{11} < \frac{3}{4}$

Only choice D provides the right order.

4) Choice E is correct

$6^4 = 6 \times 6 \times 6 \times 6 = 1,296$

5) Choice E is correct

If each book weighs $\frac{1}{5}$ pound, then 1 pound = 5 books. To find the number of books in 50 pounds, simply multiply this 5 by 50: $50 \times 5 = 250$

6) Choice D is correct

Use the volume of square pyramid formula.

$V = \frac{1}{3}a^2h \Rightarrow V = \frac{1}{3}(12\ m)^2 \times 10\ m \Rightarrow V = 480\ m^3$

7) Choice D is correct

Isolate and solve for x: $\frac{2}{3}x + \frac{1}{6} = \frac{1}{3} \Rightarrow \frac{2}{3}x = \frac{1}{3} - \frac{1}{6} \Rightarrow \frac{2}{3}x = \frac{1}{6}$

Multiply both sides by the reciprocal of the coefficient of x.

$\left(\frac{3}{2}\right)\frac{2}{3}x = \frac{1}{6}\left(\frac{3}{2}\right) \Rightarrow x = \frac{3}{12} = \frac{1}{4}$

8) Choice E is correct

The area of the circle is 16π, then, its diameter is 8.

Area of a circle $= \pi r^2 = 16\pi \rightarrow r^2 = 16 \rightarrow r = 4$

Radius of the circle is 4 and diameter is twice of it, 8.

One side of the square equals to the diameter of the circle. Then:

$Area\ of\ square = side \times side = 8 \times 8 = 64$

9) Choice A is correct

The median of a set of data is the value located in the middle of the data set. Combine the two sets provided, and organize them in increasing order:

$\{1, 3, 4, 6, 8, 10, 12, 14, 15, 17\}$

Since there are 10 numbers (an even number of items) in the resulting list, the median is the average of the two middle numbers. Median $= \frac{(8+10)}{2} = 9$

10) Choice C is correct

The area of the non-shaded region is equal to the area of the bigger rectangle subtracted by the area of smaller rectangle.

Area of the bigger rectangle $= 12 \times 16 = 192$

Area of the smaller rectangle $= 10 \times 4 = 40$

Area of the non-shaded region $= 192 - 40 = 152$

11) Choice E is correct

In the figure angle A is labeled $(3x - 2)$ and it measures 37. Thus, $3x - 2 = 37$ and $3x = 39$ or $x = 13$. That means that angle B, which is labeled $(5x)$, must measure $5 \times 13 = 65$. Since the three angles of a triangle must add up to 180,

$37 + 65 + y - 8 = 180$, then: $y + 94 = 180 \rightarrow y = 180 - 94 = 86$

12) Choice E is correct

Let x be total number of cards in the box, then number of red cards is: $x - 246$

The probability of choosing a red card is one third. Then: $probability = \frac{1}{3} = \frac{x-246}{x}$

Use cross multiplication to solve for x.

$x \times 1 = 3(x - 246) \rightarrow x = 3x - 738 \rightarrow 2x = 738 \rightarrow x = 369$

13) Choice C is correct

y is the intersection of the three circles. Therefore, it must be odd (from circle A), negative (from circle B), and multiple of 5 (from circle C).

From the choices provided, only -5 is odd, negative and multiple of 5.

14) Choice A is correct

An obtuse angle is an angle of greater than 90 degrees and less than 180 degrees. Only choice A is an obtuse angle.

15) Choice B is correct

Recall that the formula for the average is: $Average = \frac{sum\ of\ data}{number\ of\ data}$

First, compute the total weight of all balls in the basket: $25\ g = \frac{total\ weight}{20\ balls}$

$total\ weight = 25\ g \times 20 \rightarrow total\ weight = 500\ g$.

Next, find the total weight of the 5 largest marbles:

$40\ g = \frac{total\ weight}{5\ balls} \rightarrow total\ weight = 40\ g \times 5 \rightarrow total\ weight = 200\ g$

The total weight of the heaviest balls is $200\ g$. Then, the total weight of the remaining 15 balls is $300\ g : 500\ g - 200\ g = 300\ g$.

The average weight of the remaining balls: $Average = \frac{300\ g}{15\ balls} = 20\ g$ per ball

16) Choice A is correct

In the stadium the ratio of home fans to visiting fans in a crowd is $5:7$. Therefore, total number of fans must be divisible by $12: 5 + 7 = 12$.

Let's review the choices:

A. $12,324 \rightarrow 12,324 \div 12 = 1,027$

B. $42,326 \rightarrow 42,326 \div 12 = 3,527.1\overline{6}$

C. $44,566 \rightarrow 44,566 \div 12 = 3,713.8\overline{3}$

D. $66,812 \rightarrow 66,812 \div 12 = 5,567.\overline{6}$

E. $69,752 \rightarrow 69,752 \div 12 = 5,812.\overline{6}$

Only choice A when divided by 12 results a whole number.

17) Choice D is correct

Fist convert mixed numbers to fractions: $2\frac{2}{3} - 1\frac{5}{6} = 2\frac{4}{6} - 1\frac{5}{6} = \frac{16}{6} - \frac{11}{6} = \frac{5}{6}$

18) Choice E is correct

$x = \frac{1}{3}$ and $y = \frac{9}{21}$, substitute the values of x and y in the expression and simplify:

$\frac{1}{x} \div \frac{y}{3} \to \frac{1}{\frac{1}{3}} \div \frac{\frac{9}{21}}{3} \to \frac{1}{\frac{1}{3}} = 3$ and $\frac{\frac{9}{21}}{3} = \frac{9}{63} = \frac{1}{7}$. Then: $\frac{1}{\frac{1}{3}} \div \frac{\frac{9}{21}}{3} = 3 \div \frac{1}{7} = 3 \times 7 = 21$

19) Choice D is correct

Let E age of Ella, we know Ella is 4 years older than Ava: $E = 4 + A \to A = S - 3$

20) Choice E is correct

Let x be the number of current stamps in the collection. Then:

$$\frac{6}{5}x - x = 100 \to \frac{1}{5}x = 100 \to x = 500$$

50% more of 500 is: $500 + 0.50 \times 500 = 500 + 250 = 750$.

21) Choice C is correct

The sum of 8 numbers is greater than 240 and less than 320. Then, the average of the 8 numbers must be greater than 30 and less than 40.

$\frac{240}{8} < x < \frac{320}{8} \to 30 < x < 40$

The only choice that is between 30 and 40 is 35.

22) Choice B is correct

The angles on a straight line add up to 180 degrees. Then: $x + 25 + y + 2x + y = 180$

Then, $3x + 2y = 180 - 25 \to 3(35) + 2y = 155 \to 2y = 155 - 105 \to 2y = 50 \to y = 25$

23) Choice C is correct

Let L be the length of the rectangular and W be the width of the rectangular. Then, $L = 4W + 3$

The perimeter of the rectangle is 36 meters.

Therefore: $2L + 2W = 36, L + W = 18$

Replace the value of L from the first equation into the second equation and solve for W: $(4W + 3) + W = 18 \to 5W + 3 = 18 \to 5W = 15 \to W = 3$

The width of the rectangle is 3 meters and its length is:
$$L = 4W + 3 = 4(3) + 3 = 15$$

The area of the rectangle is: $Length \times Width = 3 \times 15 = 45$

24) Choice D is correct

First, find the measure of angle RQS. Angles RQS and PQR are supplementary and therefore their sum is 180 degrees. Then:

$PQR + RQS = 180 \to 135 + RQS = 180 \to RQS = 45$

The sum of all angles in a triangle is 180 degrees. Then:

$45 + 52 + x = 180 \to 97 + x = 180 \to x = 83$

25) Choice E is correct

The sum of two supplementary angles is 180 degrees. Then:

$(3x - 10) + (x + 2) = 180$. Simplify and solve for x: $(3x - 10) + (x + 2) = 180 \to$

$4x - 8 = 180 \to 4x = 180 + 8 \to 4x = 188 \to x = 47$

26) Choice B is correct

To find the area of the shaded region subtract the area of the smaller circle from bigger circle.

$S_{bigger} - S_{smaller} = \pi(r_{bigger})^2 - \pi(r_{smaller})^2 \Rightarrow$

$S_{bigger} - S_{smaller} = \pi(6)^2 - \pi(4)^2 \Rightarrow 36\pi - 16\pi = 20\pi \ in^2$

27) Choice D is correct

$$\frac{a+b}{2} = \frac{a}{2} + \frac{b}{2}$$

28) Choice E is correct

$2.5\% \ of \ 1{,}200 = \frac{2.5}{100} \times 1{,}200 = 30$

29) Choice D is correct

Solve for x: $x^3 + 18 = 130 \to x^3 = 112$

Let's review the choices.

A. 1 and 2 $1^3 = 1$ and $2^3 = 8$, 112 is not between these two numbers.

B. 2 and 3 $2^3 = 8$ and $3^3 = 27$, 112 is not between these two numbers.

C. 3 and 4 $3^3 = 27$ and $4^3 = 64$, 112 is not between these two numbers.

D. 4 and 5 $4^3 = 64$ and $5^3 = 125$, 112 is between these two numbers.

E. 5 and 6 $5^3 = 125$ and $6^3 = 126$, 112 is not between these two numbers.

30) Choice B is correct

15 second is one fourth of a minute. One fourth of 72 is 18. $72 \div 4 = 18$. Jack types 18 words in 15 seconds.

31) Choice D is correct

The amount of money that Jack earns for one hour: $\frac{\$616}{44} = \$14$

A number of additional hours that he works to make enough money is: $\frac{\$826 - \$616}{1.5 \times \$14} = 10$

Number of total hours is: $44 + 10 = 54$

32) Choice E is correct

In scientific notation all numbers are written in the form of: $m \times 10^n$, where m is between 1 and 10. To find an equivalent value of 0.000,000,000,000,042,121, move the decimal point to the right so that you have a number that is between 1 and 10. Then: 4.2121. Now, determine how many places the decimal moved in step 1, then put it as the power of 10. We moved the decimal point 14 places. Then: 10^{-14} when the decimal moved to the right, the exponent is negative.

Then: $0.000,000,000,000,042,121 = 4.2121 \times 10^{-14}$

33) Choice C is correct

A. $|4 - 2| = |2| = 2$

B. $|2 - 4| = |-2| = 2$

C. $|-2 - 4| = |-6| = 6$

D. $|2 - 4| - |4 - 2| = |2| - |2| = 2 - 2 = 0$

E. $|2 - 4| + |4 - 2| = |-2| + |2| = 2 + 2 = 4$

Choice C is the largest number.

34) Choice D is correct

85% of 40 is: $0.85 \times 40 = 34$. So, the student solves 34 questions correctly.

35) Choice B is correct

Use simple interest formula:

$I = prt$ ($I = interest, p = principal, r = rate, t = time$)

Simple interest $I = 2,500 \times 0.08 \times 6 = 1,200$

She will pay $1,200 interest at the end of 6 years.

36) Choice C is correct

Since integer x is evenly divisible by 4, substitute 4 for x in the answer choices to determine which expression is also divisible by 4: Let $x = 4$.

Choice A: $x + 1 = 4 + 1 = 5$ This is NOT divisible by 4.
Choice B: $2x + 1 = 2(4) + 1 = 9$ This is NOT divisible by 4.
Choice C: $2x + 4 = 2(4) + 4 = 12$ This is divisible by 4.
Choice D: $3x + 2 = 3(4) + 2 = 14$ This is NOT divisible by 4.
Choice E: $4x + 1 = 4(4) + 1 = 17$ This is NOT divisible by 4.

So, choice C is correct.

37) Choice E is correct

Fist, find the length of the ribbon around the box, then add 10% of it.

$(8 \times 5) + (2 \times 20) + (6 \times 10) = 40 + 40 + 60 = 140 \ in$

10% of 140: $\frac{10}{100} \times 140 = \frac{1,400}{100} = 14 \ in$, $140 \ in + 14 \ in = 154 \ in$

38) Choice E is correct

There are currently 16 balls in the bag (5 + 8 + 3). Of those balls, 11 are not blue. So, the probability of choosing a ball that is not blue is $\frac{11}{16}$.

39) Choice D is correct

First, calculate Jack's riding time in minutes: 1 hour 20 minutes = 80 minutes

Then, convert kilometers to meters: 160 kilometers = 160,000 meters

Now simplify the ratio to find the answer: $\frac{160,000}{80} = 2,000$ meters

40) Choice E is correct

If x is the smaller consecutive integer, then $x + 1$ is the larger consecutive integer. Use their sum (-13) to find x:

$x + (x + 1) = -13 \to 2x + 1 = -13 \to 2x = -14 \to x = -7$

The two consecutive integers are -7 and -6. 2 is added to the smaller integer: $-7 + 2 = -5$, and 3 is subtracted from the larger integer: $-6 - 3 = -9$ find the product: $-5 \times (-9) = 45$

41) Choice D is correct

The two greatest integers less than -3.34 are -4 and -5. Since -5 is odd, the answer is -4.

42) Choice A is correct

$$(2)^3 + (-3)^2 + 2x - 6 = 11 \to 8 + 9 + 2x - 6 = 11$$

Combine like terms: $11 + 2x = 11 \to 2x = 0 \to x = 0$

43) Choice D is correct

The relationship among all sides of special right triangle $30° - 60° - 90°$ is provided in this triangle:

In this triangle, the opposite side of $30°$ angle is half of the hypotenuse.

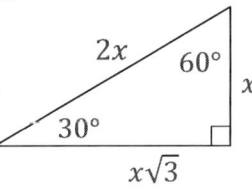

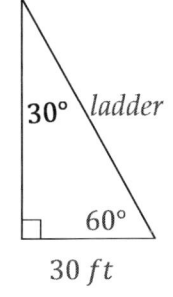

Draw the shape of this question:

The ladder is the hypotenuse. Therefore, the ladder is $60\ ft$.

44) Choice B is correct

$(6 - 1) \times 160 = 800$ centimeters

45) Choice D is correct

Plug in the values of x in the choices provided and check the answers.

A. $y = \frac{2}{3}x \to y = \frac{2}{3} \times (2) = \frac{2 \times 2}{3} = \frac{4}{3}$

B. $y = \frac{3}{2}x \to y = \frac{3}{2} \times (2) = \frac{3 \times 2}{2} = \frac{6}{2} = 3$

EffortlessMath.com

C. $y = 2(x - 1) \rightarrow y = 3(2 - 2) = 3(0) = 0$

D. $y = 3x - 1 \rightarrow y = 3(2) - 1 = 5$

E. $y = 4x - 2 \rightarrow y = 4(2) - 2 = 6$

Only Choice D is correct.

46) Choice D is correct

Let's find the mean (average), mode and median of the number of cities for each type of pollution. Number of cities for each type of pollution: 6, 3, 4, 9, 8

$average\ (mean) = \frac{sum\ of\ terms}{number\ of\ terms} = \frac{6+3+4+9+8}{5} = \frac{30}{5} = 6$

The Median is the number in the middle. To find median, first list numbers in order from smallest to largest. 3, 4, 6, 8, 9. The median of the data is 6.

Mode is the number which appears most often in a set of numbers. Therefore, there is no mode in the set of numbers. Median = Mean, then, $a = c$

47) Choice A is correct

Percent of cities in the type of pollution A: $\frac{6}{10} \times 100 = 60\%$

Percent of cities in the type of pollution C: $\frac{4}{10} \times 100 = 40\%$

Percent of cities in the type of pollution D: $\frac{9}{10} \times 100 = 90\%$

48) Choice A is correct

Let the number of cities should be added to type of pollutions B be x. Then:

$\frac{x+3}{8} = 0.625 \rightarrow x + 3 = 8 \times 0.625 \rightarrow x + 3 = 5 \rightarrow x = 2$

49) Choice C is correct

Let x be the missing number. Then:

$\$365.76 + \$289.55 + \$378.45 + \$289.99 + \$1,459 + x = \$3,332.49 \rightarrow$

$\$2,782.75 + x = \$3,332.49 \rightarrow x = \$3,332.49 - \$2,782.75 = \$549.74$

50) Choice B is correct

Use the average formula:

$average\ (mean) = \frac{sum\ of\ terms}{number\ of\ terms} = \frac{9+12+15+16+19+16+14.5}{7} = 14.5$

Effortless Math's CBEST Online Center

... So Much More Online!

Effortless Math Online CBEST Math Center offers a complete study program, including the following:

- ✓ Step-by-step instructions on how to prepare for the CBEST Math test
- ✓ Numerous CBEST Math worksheets to help you measure your math skills
- ✓ Complete list of CBEST Math formulas
- ✓ Video lessons for CBEST Math topics
- ✓ Full-length CBEST Math practice tests
- ✓ And much more...

No Registration Required.

Visit **EffortlessMath.com/CBEST** to find your online CBEST Math resources.

The Best CBEST Math Books!

Download eBooks (in PDF format) Instantly!

Most Popular CBEST Math Books!

Receive the PDF version of this book or get another FREE book!

Thank you for using our Book!

Do you LOVE this book?

Then, you can get the PDF version of this book or another book absolutely FREE!

Please email us at:

info@EffortlessMath.com

for details.

Author's Final Note

I hope you enjoyed reading this book. You've made it through the book! Great job!

First of all, thank you for purchasing this study guide. I know you could have picked any number of books to help you prepare for your CBEST Math test, but you picked this book and for that I am extremely grateful.

It took me years to write this study guide for the CBEST Math because I wanted to prepare a comprehensive CBEST Math study guide to help test takers make the most effective use of their valuable time while preparing for the test.

After teaching and tutoring math courses for over a decade, I've gathered my personal notes and lessons to develop this study guide. It is my greatest hope that the lessons in this book could help you prepare for your test successfully.

If you have any questions, please contact me at reza@effortlessmath.com and I will be glad to assist. Your feedback will help me to greatly improve the quality of my books in the future and make this book even better. Furthermore, I expect that I have made a few minor errors somewhere in this study guide. If you think this to be the case, please let me know so I can fix the issue as soon as possible.

If you enjoyed this book and found some benefit in reading this, I'd like to hear from you and hope that you could take a quick minute to post a review on the book's Amazon page.

I personally go over every single review, to make sure my books really are reaching out and helping students and test takers. Please help me help CBEST Math test takers, by leaving a review!

I wish you all the best in your future success!

Reza Nazari

Math teacher and author

Made in the USA
Las Vegas, NV
12 March 2025